DOING BIRD

A Chicken Keeper's Year

Martin Gurdon

Constable • London

Constable & Robinson Ltd
55–56 Russell Square
London WC1B 4HP
www.constablerobinson.com

First published in the UK by Constable,
an imprint of Constable & Robinson Ltd., 2013

A copy of the British Library Cataloguing in
Publication data is available from the British Library

ISBN 978-1-78033-193-5 (paperback)
ISBN 978-1-78033-398-4 (ebook)

Printed and bound in the UK

1 3 5 7 9 10 8 6 4 2

Acknowledgements

Special thanks to Sheila Ableman, whose hard work made this book happen, and whose thoughtful advice made it better. Leo Hollis, Charlotte Macdonald, Sue Viccars, Andreas Campomar and everyone at Constable & Robinson for more of the same. Elvi Murcell (I nicked something she said about her hair), David and Jenny Gurdon and especially my wife Jane.

For Philip Gurdon, who liked words too.

PREFACE

Welcome to the sometimes-weird world of the amateur domestic bird keeper.

I'm a writer who has spent the last fourteen years sharing a rural Kentish back garden with my wife, a dog, a cat and an ever-changing flock of chickens. More recently, this space has grown busier, with the arrival of Indian Runner ducks and a fluctuating colony of doves. The surprising, sometimes moving and often funny ways they reacted to us, and each other, lie at the heart of this book. Charting their arrivals, lives, enthusiasms, fears and exits also drives the story it contains. If you've thought about keeping hens or other domestic birds – as thousands of people have in recent years – I hope *Doing Bird* will show you what this can be like; and, if you're there already, that you'll get a feeling of recognition.

This book covers the story of a year spent as a bird keeper in a domestic setting: the social and seasonal expansions, contractions, comings and goings that the human and animal occupants of our home experienced over that time. Some episodes are based on events that happened in earlier years; sometimes I've ferreted about in our recent history to expand on them. This means that some of the birds or animals we've owned, but who are no longer with us, put in the odd appearance too.

Ultimately, writing *Doing Bird* felt like taking a trip over the recent past, and I enjoyed the journey. I hope you do too.

WINTER

CHILLING OUT

My hand was stuck.

It was 5.30 on a winter weekday morning, and the sky was a cloud-free deep blue, lit by a big full moon. As I'd made my way through the garden I hadn't needed to use my torch. I could see my muddy green wellingtons sinking ankle-deep into the moonlit, meringue-hard snow, which cracked with each step. Then I tripped, and nearly fell face first into the snow, instead breaking my fall with gloveless hands.

Cursing softly, I approached the henhouse. Everything was very still, cloaked by a pervasive cold. Naked, wet hands and a sub-zero temperature made touching things a dangerous proposition, but I still had to unlock one of the aviary doors, secured by a cheap metal latch which felt as if it was made of metallic marshmallow. This was so bloody cold the skin of my numbed fingers instantly stuck to it: Nature's way of saying 'Go away, dry your hands and put your gloves on.' My brain was functioning on reflex and I was deaf to Nature's entreaties, so wrenching my hand free from the latch I opened the door and grabbed the chicken drinker.

Looking like a child's drawing of a Soviet space rocket, this stood about two feet tall, and was made in two parts: a removable cylinder with a handle and a pointy top, which covered another cylinder. The latter was attached to a sort of dish arrangement from which the chickens drank. Once it was exposed you poured water into it (filling the dish), then refitted the outer cylinder, a bit like a giant robot's galvanized condom.

On cold mornings like this one the drinker only worked if you'd emptied it the night before. Once again I hadn't and the contents had frozen solid, making opening and refilling the accursed item impossible. The handle was bent sideways, thanks to a previous unsuccessful attempt to shatter the frozen contents by turning the drinker upside down and banging the lid on the rock-hard ground.

As I wrapped my already numbed hand round the handle it instantly felt as if my icy fingers had been welded there. Unable to move them, I was now bent in a lumpy arc over the drinker. Frosted blasphemy broke the silence as I contemplated my rapidly freezing hand, uselessly gripping its handle.

Fearing that prising it off would result in removing a layer of skin, I considered carrying it back to the house, where we could both warm up enough to be parted without injury. This seemed like a good plan, but when I tried straightening up I found that the drinker had also welded itself to the ground. Muttering 'Shit', I kicked it, but it didn't budge. With an escalating feeling of panic

4

I kept kicking until one of my boots crunched against the drinker with a dull clang, viciously compressing the toes of my right foot. I realized that normally this would have hurt a lot; in fact, had things been a little warmer, and my circulation working properly, it would have been bloody excruciating. Instead the biggest toes of my right foot just felt deader than before.

Releasing another stream of invective I gave the thing a final, brutal kick, which dislodged it from the ground, leaving a dinner-plate-sized disc of frost-free earth. Now I could stand up straight, but was still fused to the drinker's handle. Should I go back into the house like this? An absurd vision of clanking into the bedroom, cup of tea for my wife in one hand and chicken drinker stuck to the other, floated into my mind. Sod it; I was just going to have to prise it off my fingers, and live with the consequences. Clasping my frozen hand, I yanked downwards, feeling the flesh stretching until, suddenly, I was free, and apparently not in agony. Since I'd expected it to hurt a lot I'd made a great deal of noise, and from inside the henhouse there could be heard some avian shuffling, followed by irritable *'Don't you know what time it is?'* clucking.

Our birds roost in a garden shed with a saggy floor, on either side of which is a pair of big, home-made aviaries with doors secured by catches made from cheap, soft metal. Both aviaries contain garden benches that have become too wobbly for human bottoms to sit on, but are capable of sustaining a perching Buff Orpington.

Normally our flock of chickens is more than keen to get up. Hearing me coming will result in a lot of argy-bargy and milling about in the dark, as the girls collide with each other or perhaps engage in some low-key bullying. Sometimes Svenson the cockerel will make his presence felt by crowing in a determined macho fashion, especially if he hears me stamping about and decides that I've been too slow in letting everyone out.

I opened a trap door to release the birds, and began pouring their breakfast into the feeder. Eventually something feathery appeared, peered up at me briefly, and then withdrew. I took a moment to inspect my hand by torchlight, and saw a slightly alarming pale yet liverish-looking mark, running diagonally along my palm, as if it had been on the receiving end of a ruler-wielding, cut-price suburban dominatrix. My wife would understand, but it might be difficult to explain to others. I flexed my fingers to try and warm things up a bit, then rubbed my hands together, hoping the friction would agitate their frozen nerve endings into some sort of activity. On this level it was successful, but I sensed that the tingling I was starting to feel was a precursor to pain. I knew that I wouldn't be brave about it when it arrived.

To take my mind off this I rattled the feeder. The head in the henhouse doorway appeared again and Slasher looked up at me without obvious enthusiasm, so I banged down the lid of the dustbin where we keep the chicken's food and noticed a skittering movement, as a small, pale-brown shape made a businesslike dart for the feeder. After huddling inside the henhouse, Slasher had decided to ignore the icy conditions and give in to greed. One of our younger birds, Slasher is a compact, busy little chicken with mottled brown plumage, a pompom of feathers erupting from the top of her head and a red comb, like a Renaissance artist's hat, flopping rakishly over one eye. She began pecking enthusiastically, causing the feeder to sway and wobble on the end of the domestic flex from which it hung.

Now there was a lot of avian muttering and shuffling from inside the henhouse, but still nobody else was joining in. Slasher made the best of the opportunity for an uninterrupted feed, but this state of affairs wasn't going to last. For a chicken, hearing another chicken filling her face when you could be filling yours – and in the process preventing her from doing so – is something which won't be endured for long. So Ulrika, one of our bossier birds, emerged from beneath the trap door and headed for the food, shoving Slasher out of the way. Ulrika's Dad was called Sven, on account of his alleged Swedish origins. Her mother was a Welsummer hen named Ann, another figure of authority. In the henhouse primacy is the order of the day, and Ulrika was clearly a member of the ruling class. She had the brown plumage of a regular farm hen, but was a little more athletic; an odd word to describe a chicken, perhaps, but it suited Ulrika, who looked a bit like a game bird.

She also looked hard, because she was. If Ulrika had been a teenage schoolgirl her hair would be hauled back from her forehead in a vicious ponytail, and she'd probably regard kickboxing as the only after-school activity worth bothering with. And if chickens carried flick knifes, Ulrika would have been armed.

So when she gave Slasher the eye – a process that involved placing her face very close to the other bird and staring hard – Slasher backed off. Ulrika got stuck in and soon the machine-gun rattle of her beak against the plastic dish resulted in some of its contents spraying round the icy run. Slasher quickly made the best of Ulrika's rotten table manners by pecking up her leftovers.

Slasher's moniker is also a nod to her violent inclinations, and came to us soon after we'd bought her, along with another chicken. Lacking inspiration on what to call these birds, we'd temporarily christened them 'the Other Two'. We've always given our hens names, something which many of our hardcore rural friends think is inexplicable and moderately nauseating. We avoid cuteness wherever possible, so there are no Tulips and Buntys in our flock. Bad taste is the order of the day, and hens that have come our way over the years have included Satan, Bald Bird, Peckham and Egghead.

For us, the rules of chicken naming are simple. No pun is too awful; no alliteration off-limits; and often we simply wait to see what the birds are like, and for the right names to suggest themselves. So it was with Slasher. One evening when we were going out I had to round up the hens and confine them to the run. Slasher is a free spirit who doesn't take kindly to being herded and imprisoned, so this was a process that first involved catching her in the garden.

I'd had a lucky break and taken her by surprise, scooping her off the ground, causing her to emit a terrible scream – which my wife Jane heard in the house and thought was the result of the chicken being attacked by a fox. Wriggling sinuously, the bird tried twisting herself free, but when that didn't work she turned and grabbed hold of the flap of skin between my thumb and forefinger and began wrenching it about. This didn't actually hurt, but it did demonstrate a certain strength of character. When I explained to my wife what had happened, describing the bird as 'a little slasher', we realized that she had all but named herself.

This left us with the problem of finding a name for the other hen. A petite, beady-eyed Welsummer crossbreed with dark feathers interlaced with neat, almost gold, geometric patterns, she remained the nearest thing to a regular companion Slasher had. Being the anonymous half of what had been 'the Other Two' meant that this bird's moniker eventually became shortened to 'Too'. If this is an expedient name, then the hen that bears it is anything but. Like Slasher, Too is a contrary creature, and almost completely fearless. She certainly doesn't find me intimidating. So as the grey dawn light began dribbling into the sky like milk in weak tea, and the serious business of feeding at last got underway, she emerged from the henhouse, jumped on my boots, and began eyeballing me.

Too loves bread, and some mornings I dole a little out as an avian treat. For the hens we buy cheapish ready sliced, but for the sake of their health it's brown. Where possible we go for the higher-quality stuff when it's reaching its 'sell by' date and being sold off at knock down prices by the local supermarkets. We think the cheapest own-brand white (which appears to be made from freeze-dried asbestos) isn't good for chickens, let alone people. The bread lives in the metal food bin as well as the bags of grain and layers pellets which are the staples of our birds' diets. Too will always be one of the first in the scrum to get her share.

Her determination to feed sometimes got Too into trouble. She often became so excited that she perched precariously on the rim of the food bin, flapping her wings to keep her balance. More than once she'd fallen in. One morning when everyone was especially keen to shovel the stuff down their beaks and there was a lot of jostling, I plonked the food bin lid on the ground and started distributing largesse. After a bit I became aware that Too wasn't among the takers, which was thoroughly out of character. Then I noticed the dustbin lid. It appeared to be moving. Lifting it like a galvanized terrene revealed a slightly dishevelled chicken, who was so traumatized by the event that she launched herself into the air, yanked a piece of bread from my free hand and shot off with it.

As she, Ulrika and Slasher got stuck in, the others drifted in from the henhouse and joined them. Squawks was the bird at the very bottom of our flock's pecking order. A large, round, black, feathered pompom, she's a strange combination of idiocy, hysteria, ungainliness and beauty. You could half imagine an American football cheerleader waving her about. Her beak and legs are as black as her feathers, and she has a very peculiar run. At low speeds she doesn't quite limp, but her gait is uneven, and when she's hurrying it becomes a demented up-and-down lollop. When she thinks you might feed her, she will erupt into view and boing! boing! boing! towards you like a feathery bouncing bomb.

What makes her beautiful? Her eyes: coal-black, feminine circles, fathomless pools of not very much. With the best will in the world, Squawks is a twit; a nice twit, but even by the standards of the average chicken she's thick, the avian equivalent of a dim debutante. Think minor ex-royal without the financial acumen, but with added avian dyspraxia. Squawks is the clumsiest hen ever to have crossed our henhouse threshold. She lurches about, cannoning into the others, who will often express their displeasure by giving her a serious pecking. She has a penchant for tangling herself up in foliage, and will struggle from the undergrowth, dragging large twigs or small bits of tree behind her, as she hauls herself towards anything edible.

Being at the bottom of the heap means that she has to work harder than the others to make sure she's fed, and it's taken her a while to develop the strategies needed to guarantee this. Originally she simply charged into the twice-daily feeding mêlée, with predictably unhappy consequences. Now she will hang back a little, looking for opportunities to make lumbering surgical strikes on the grub. If it's bread, which she loves, she'll charge off with her prize before anyone else pinches it.

Such behaviour would never occur to Brahms the Brahma. Now seven, she is one of our oldest hens, so should be near the top of the pecking order. She has a wizened red face, hooked beak and a rolling, ponderous gait. Her plumage is almost a Paisley pattern of black lines on dark-brown feathers, an incongruous contrast with her solid physique and tough-nut visage; as is the fact that she is remarkably unassertive. Younger birds have emerged from eggs or arrived in cardboard boxes then climbed up the chicken social ladder, more or less trampling on Brahms as they ascended. We've never been entirely sure why, but putting it crudely, for the cockerels in her life, Brahms was never a particularly good lay, in any sense of the word.

She gave up laying eggs after a couple of years, and hasn't shown the slightest inclination to go broody and hatch out anyone else's. In the unreconstructed world of the henhouse these things represent elements of desirability that push you up or down the popularity stakes. Fecundity gets you to the front of the queue at mealtimes and sometimes puts you in charge of some, or all, of the other birds. So the eggless Brahms also initially kept her distance when the food arrived, and made her way to it after the more senior hens had taken what they wanted. I often keep an eye on her to make sure she isn't chased away and so goes without completely.

Bonnie is an example of a younger chicken who arrived after Brahms and is now her senior. A white, speckled affair of indeterminate parentage, Bella came to our household as an egg. We named her after one of the pubs in our village. She used to lay a lot of eggs but has largely given this up in recent years, and after having one attempt at parenthood by going broody and spending an unsatisfactory month sitting on some duck eggs and failing to hatch anything, hasn't repeated the experience.

Now pretty near to the top of the pecking order, she approaches life like a steely middle manager in the sort of work environment where the term 'creative tension' is used to describe bullying. She can be benign – or at least about as friendly as a chicken gets – then will suddenly give poor old Slasher, Too, Squawks or Brahms a quick, vicious seeing to, involving the yanking out of some neck or back feathers. This doesn't always happen at mealtimes, but is quite likely to. This particular morning Bonnie just gave the lower ranks a special, basilisk stare before helping herself.

Bonnie is not to be trifled with, something Meringue, our Marans hen, found to her cost when she tried to bully her. If the word 'Marans' looks a bit odd, that's because it's French. Meringue is a precise, speckled grey bird who makes distinctive, almost keening noises, and would like to be forceful too. A contemporary of Brahms, she also ought to be a lot further up the pecking order. To prove her credentials she once squared up with Bella for no obvious reason, coming off second-best from the skirmish, retiring if not hurt, then a bit dishevelled. Peace appeared to be restored, until several days later when she had another go, lost again as badly as before, and was comprehensively roughed up. This was a process involving a lot of dusty dragging about and futile attempts to separate them.

Chickens learn by experience, but sometimes need to have that experience several times for it to properly sink into their little brains. So it was with Meringue, who had more goes at Bonnie before giving up, pained if not actually injured; after that the violence stopped, and now they seem to get on fine. These days Meringue is most likely to be found in the company of Brahms, with whom she seems to have an understanding, despite being poles apart socially, and they'll trundle around the garden like a pair of old ladies on a trip to a village hall coffee morning. Breakfast, however cold, is a different matter, and this morning, as Brahms hung back, Meringue shouldered her way into the freezing, feeding pack.

If Brahms and Meringue vaguely resembled extras in an episode of *Miss Marple*, Ann Summers, Ulrika's mother, was more Rosa Klebb with feathers. Ann was a bird with a Lotte Lenya 'my-shoes-have-blades' strut. This was partly because she had not long assumed the role of chief chicken, so had nothing to prove. She was power-hungry by inclination, born to make others do her bidding, and rarely needing to resort to force. She was also a serial parent, and went broody during most of the summers she'd been with us. That's how she arrived, hot and hormonal, from a lady chicken breeder in the Weald of Kent. She was mother to both Ulrika and our current cockerel Svenson, and stepmother to Bella, who came to us as an egg from another flock. Ann sat on the egg and hatched her out.

Svenson and Ulrika's Dad, Sven, is no longer with us, but turned out to be rather a good thing as cockerels go. On the move he had a stately gait, and resembled, just slightly, a Prussian army hat on legs. He was solicitous to all his girls, not too brutal when it came to sex, not prone to attacking people and distinctly elderly when parenthood with Ann finally happened. Two boys and a girl resulted. His sons lacked Sven's gravitas, but made up for this with a riot of burnished copper and deep-green plumage. Their sister Ulrika had more than inherited her mother's spiky demeanour, and quickly displayed a sense of entitlement, which apparently comes from being part of the right bloodline.

This power play was in full swing when Sven keeled over, and we decided to keep one of the cockerels, christened Svenson (because that's who he was). His brother went to friends in the village where we used to live when we first started keeping chickens. Their daughter named him Swede-y Todd (although confusingly he's also known as Sven). Swede-y Todd/Sven 2 soon proved his credentials as a babe magnet, and rapidly cut a swathe through our friends' flock of lady chickens, who inevitably sired another bunch of cockerels to challenge their Dad and make a load of noise. Mercifully, like his father and brother, he hasn't been violent towards people.

Svenson himself, who was now in his third year, had grown into a big, handsome animal with an equally keen interest in procreation, and a sense of girlfriend favouritism that Meringue in particular has found more than trying. He won't leave her alone, but had rather less to show for it than his brother. Chickens are not given to angst, and as Svenson strutted his stuff, indicating to his chosen ladies where he thought the best feeding places were, he did not seem to be troubled by the lack of an heir. After thirty-six months he really ought to have known what he was doing but it's painfully obvious to everyone that Svenson is still getting it wrong, but more than willing to keep practising.

Frankly he has problems angling bits of his wedding tackle. During moments of avian passion he looks like a jockey on the final furlong, arse up in the air, with little hope of a successful coupling. During these extended demonstrations of inept bum waggling his unwilling amours remain squashed in the long grass looking variously alarmed, irritated or resigned. Then, once the action has finished, they bolt for freedom. That winter's morning it was sufficiently cold for Svenson to keep his bits to himself, and he seemed contented with a little cockerel war dance (drop one wing, shuffle sideways whilst jiggling your feet up and down as if you're kneading bread very quickly), then got on with the serious business of feeding.

Bolting for freedom was by now pretty much my plan as well, as I was keen to take my throbbing hand indoors to thaw it (and the rest of me) out.

First I had to do something about getting the chickens some water. In the short term I would fill up an old plastic washing-up bowl, and take their proper drinker into the house to thaw out. That seemed like a good plan, so for the second time that morning, I picked it up with a gloveless hand.

My wife swears that at the same time she was woken by a voice coming from the garden. It was shouting 'Bugger!'

*****

'It isn't going to hurt,' said Jane.

'But that water will be cold,' I whined.

'Not as cold as your hand,' said my wife. 'So the water will probably feel warm, and hot water really will hurt.'

We were standing in the kitchen. My wife, who was swathed in a dressing gown and slippers, looked sleepy, her copper-coloured hair a little wild. I was still dressed for the Arctic. Even my hood was still up, so that I looked like a Siberian train-spotter, except that I was dangling a dirty metal chicken drinker over the sink rather than clutching a notebook. Jane had a jug of water, which she wanted to pour over my hand to help free it. The prospect had released my inner toddler, who was now having a serious grizzle.

'I've been bloody stupid.'

'Yes.'

'And do you know what's worse?'

'You getting me out of bed with this?' asked Jane, gesturing at the chicken drinker with her jug.

'No.'

Noticing my wife's sudden coolness, I muttered, 'Well yes, obviously that too, but it's not just that, it's my other bloody hand. I've picked this thing up with both of them now, and the first one's throbbing, so this is going to be painful as well.'

My wife has large, expressive eyes. She used them to give me a stern look.

'You should have worn gloves.'

'I know that,' I said irritably, 'but . . .'

'You didn't,' cut in Jane briskly. She has developed a nifty line in finishing off my sentences, which is especially annoying when she's right.

I flinched as she began pouring water on to my hand.

'It doesn't hurt,' said my wife, measuring each word slowly and deliberately.

She was right about that too. The water almost felt pleasant, and I began flexing and rolling my handle-trapped fingers, until I started feeling the skin pulling away from the metal. Although there was no pain (that would come later) it still felt remarkably unpleasant, as if the undersides of my fingers were being sucked from their bones by a vacuum cleaner, which at the same time was blasting them with sand grains. As I finally prised them free I pulled another face, and Jane told me not to make a fuss.

'You can make me some tea instead, and take that filthy thing out of the sink,' she said brightly, pointing at the drinker, before heading out of the kitchen, dressing gown swishing. As I watched her go I knew that I was now a debtor on life's balance sheet.

Wrapping some paper towel round the drinker's handle I hauled it out, dumped it on the kitchen's stone floor, and having extracted dried soil and dead grass from the plughole looked ruefully at my palms, which now both sported the same peculiar striped marks. My right hand continued to feel uncomfortable. A few minutes later, as I climbed the bedroom stairs carrying two cups of tea, the left hand joined in by throbbing too . . . but by then I had something else on my mind.

GONE WALKABOUT

'Which chicken's missing?' asked Jane as she sat up in bed and blew the steam from her teacup.

'Nude.'

Distracted by having a chicken drinker welded to my hand, it had taken a while for me to clock that we appeared to be missing a chicken. Nude was a Buff Orpington (get it?), which meant she was large and almost golden yellow, looking a little like Squawks, but

19

without the peculiar gait. Some Buffs are blowsy and laid back to an almost hippyish degree, but Nude, who arrived in our garden as a pullet – which is an adolescent bird, no longer cute and babyish but still growing up, rather like the average human teen – soon proved more strident. She had 'survivor' written all over her, and quickly established herself in our flock as a bird who wasn't about to be oppressed by older, tougher hens. Sitting somewhere in the middle of its social strata, she was neither put-upon runt nor boss bird.

Keep hens for long enough and you will quickly conclude that they're generally pleased to see you for two reasons: protection and food. Those at the bottom of the pecking order, especially the most recent arrivals, will often shadow you, keeping a reasonable distance but making their presence obvious. They seem to understand that you'll act as security if any of their feathered cohorts tries to give them a hard time.

Everyone associates you with food, and since eating equates to survival and presumably pleasure, that really makes your company worth cultivating. Give chickens some tinned sweetcorn and they will go completely mad for it, scrapping and scrumming to get as much of this anodyne food stuff as they can. Which is odd because, apparently, they don't have taste buds.

Occasionally we feed our birds some food scraps, which cuts down domestic waste and seems to be a good way of turning stuff the chickens might enjoy into fresh eggs, but we draw the line at curry, or curried rice, on the basis that this could make the birds rocket-powered. Some chicken breeders think feeding scraps is a bad idea, as it spoils the birds and makes them reluctant to eat their regular chicken feed because they'll find it boring by comparison.

It also means that they won't get all the nutrients they need for the energy-sapping process of egg laying, and their annual feather moults, and although you don't get morbidly obese chickens they can be made ill by the wrong diet. Since ours are pets first we steer a middle course, mostly feeding them thrilling layers pellets, with occasional small rations of more exotic stuff.

But once they associate you with food of any description, you'll never walk alone. When we go near the chicken's food bin, we can usually guarantee an attentive audience. Some make sure they're not in catching distance, but others will focus on you like feather-clad hypnotists. You can be gardening, and a knot of birds will turn up and give you their full attention. Digging a hole has the potential to stir up all sorts of edible delights. Nude understood about eye contact and personal space (as in 'making it' and 'invading it'), so Jane and I quickly became used to having a rotund, fluffy stalker on our trail whenever we ventured into the garden. She also proved to be a bird who liked her creature comforts, hiding under bushes when it rained (unlike some of the other hens, who stand in the middle of the garden for hours to get the full benefit of a freezing soaking, then huddling in a damp henhouse for a quiet steam as they wait for life-threatening chest infections to kick in).

So for a chicken, Nude possessed an unusual degree of nous, and her absence was especially worrying.

'You'd better have a look for her,' said Jane. 'I'll sort out breakfast.'

'I hope I don't find a pile of feathers in the garden.'

'Well if you do, at least we'll know what happened,' said my wife.

The outside temperature had dropped to minus seven overnight. Cold enough to have finished off even a young, healthy hen. I began the search, cursing that I hadn't checked on the birds the night before, although I knew that once the previous evening's light had gone Nude would have hidden herself away. Hens are unable to see in the dark and are skilled at roosting inconspicuously. Not having a predator's finely tuned vision, sense of smell and ability to move quietly and quickly, it would have been very unlikely that I'd have found her. Still, I would have tried.

As I poked about under the brittle, wintry branches of a bush, something trod lightly on my foot, and stayed there. It was Too, whose *stand on the human's foot and give them a hard look to see if they'll give you more grub to make you go away* communication technique has become something of a party trick. We peered at each other. What she saw was a gangly man in his forties with curly hair, which once grew in such a way that in his very distant teenaged past his father referred to him as 'Bubbles', something that always resulted in a lot of humourless, adolescent huffing. Which of course was the idea. This thatch was now jammed under a shapeless woollen hat, the rest of its owner trussed up in various cold-defying bits of clothing, including a coat known as 'the Tent'.

This was once a very expensive, high-quality outdoor jacket of soft green quilted material that did a fantastic job of keeping the weather out and the heat in. It had once belonged to my Dad, had then been 'borrowed' and after twenty years never returned. I'd like to claim that this was revenge for the 'Bubbles' jibe, but it was simply theft. Time, cleaning out chickens and being worn when hacking back rampaging brambles had taken its toll on the Tent, which had grown soiled, shapeless and torn. It had become the sort of garment that could stand up on its own, but it did made me easy for a chicken to spot, and we eyed each other for a few seconds.

'I've bloody well fed you already,' I said, each word producing a cloud of vapour, 'and I've got to try finding your chum. You can wait.'

Wheezing softly, Too shat expressively next to my foot, got off it and wandered away.

'What a critic,' I muttered and went back to hunting for Nude. Half an hour later I'd found nothing, which was a mixed blessing. At least there was no sign that she had been attacked in the garden, but that didn't mean that she hadn't been unlucky somewhere else. Back in the house, Jane handed me a cup of coffee around which

I wrapped still-tingling hands while giving her a progress report.

'I'm not sure where else to look,' I said. 'At this rate I'll have to ask the neighbours if I can search their gardens.'

Like many country dwellers we started life in the town, growing up in the same bit of Edwardian south-west London, but we've gravitated to a Weald of Kent village because my wife – who is far more dynamic when it comes to home making than I am – wanted to escape her urban roots. I liked city living and still miss it sometimes, but came along for the ride, and found a lot to enjoy.

We moved to a place where there were fields rather than DIY superstores at the bottom of the road. Some of our neighbours moo and bleat at us rather than complain about the lack of on-street parking, or try to mug us. Our home isn't a farmhouse cottage with exposed beams on which to bang our heads, or priest holes into which we might fall after a drunken night out. No, it's a mid-Victorian semi-detached on the outskirts of a pretty village, where the countryside dips and undulates, but not far from the flat, slightly otherworldly landscape of Romney Marsh. Our bay-fronted home, made of dark-red brick, with its deep windows, tall rooms and broad, straight staircase, could easily pass unnoticed on Streatham Hill. It's a townhouse that lives in the country, and this familiarity was one of the reasons that I really liked it and instantly felt at home.

It wasn't Jane's idea of a rural house, but she has grown fond of it too, and its interior, with walls painted in warm reds, creams and dark greens, and its mix of furnishings, from thirties' armchairs with exotically curved wooden arms to the austerity era kitchen cupboard with frosted glass doors, ribbed cream plastic handles and Festival of Britain primary colours. We found that in a junk shop in New Cross, and it's part of a gentle aesthetic jumble of stuff we've accumulated over a long period. The end result – a sort of giant glory hole of books, trinkets and ephemera – wouldn't suit everybody, but it works for us.

Our house was originally called High View, and lives up to its name in a way few urban homes can match, because it sits on a raised section of ground, overlooking fields (at least until a developer fills the gap with dull 'executive-style' homes). From the upstairs windows it's possible to see countryside covered by still relatively small fields that sometimes slide from view beneath patches of woodland. The result is a visual jostling of greens, browns and, in the winter, near blacks of leafless trees. When it's been snowing, everything becomes uniform beneath icing-sugar layers.

We stood at an upstairs window looking down on a blank whiteness of garden and the field beyond, drank our coffees in the warm, and wondered where Nude was. The sky seemed bigger, and the early morning light, amplified by the snow, was bright and hard. We discussed chicken-rescuing tactics, including searching our neighbour Joyce's garden. Joyce – whose house is joined to ours – is a patient, practical widow. She was born and brought up in the village, and has a retinue of adult children and grandchildren living nearby. One of those hubs of village life, Joyce was used to our bird-related crises, so if I needed to poke about in her garden trying to find Nude she'd just shrug and laugh.

Once, before the Great War, our houses were linked by an interior doorway and used as a children's home, and when we moved in you could just make out its outline in the plaster. We've found no other legacy of this period, but with rooms whose shapes mirror one another, and gardens that used to be part of a single orchard, our home and Joyce's still feel joined at the hip. At the very bottom of our gardens are open fields, and I hoped strongly that Nude hadn't managed to find a way into those. If she had we'd probably never see her again. Opposite was the garden of our other neighbour, a railway track worker with a big family; but he'd temporarily moved away, having torn off the back of his house to build an extension, and his garden was stacked with building materials. It was inaccessible and filled with potential chicken hiding places. We hoped that Nude hadn't found a gap in the fence and got herself lost there.

As I mostly work from home, there would be plenty of opportunity to go back and look again during the day. My wife, who is an educational special needs advisor, wouldn't be joining the hunt; she'd be out advising.

Back downstairs, as we continued discussing the absent chicken, Mollie, our prehistoric cat, arrived. She is now the only legacy of the start of Jane's career as a teacher in the Venice-like fabulousness that is Romford, a period which involved living in an Ilford bedsit. Ancient and slightly moth-eaten, Mollie was rescued from being an Ilford street walker to live a slightly bourgeois rural existence, which initially involved embracing country sports with a prolonged spree of mass bird and shrew slaughter. Now largely retired from murdering things, Mollie contented herself with lacerating the sofa with her claws. As the stropping continued, Jane clapped her hands and shouted 'Mollie, NO!'

Being a cat, and (although we didn't realize it at the time) having gone deaf, Mollie paid no attention. Then Hoover, our knee-high crossbreed terrier dog arrived. For Mollie, Hoover is the anti-Christ, a hideous blotch on an otherwise perfect world, who has to be either bullied or avoided. This morning avoidance was her preferred tactic. She tried to flee but got a claw tangled in the frayed sofa cover, so was unable to escape as he playfully shoved a damp black nose into her backside. The hissing, thrashing outrage this provoked put an end to the chicken conversation, as we separated dog from cat, and cat from sofa cover. After that Mollie stalked off, tail up and splayed out like an enraged toilet brush.

'Where were we?' I asked as Hoover rammed his bearded face and cat-baiting nose into my crotch, inhaled, shook his head as if he was yanking a truffle from the ground, and then jumped onto my lap. After trampling my testicles he curled up, sneezed loudly and began licking my hand with a warm, abrasive tongue. As the pain subsided, I massaged the dog's ears with lightly spittled fingers, noticing that Jane was smiling.

'What are you smirking at?'

'You,' said my beloved. 'You're doing your "pompous pained" face.'

Ignoring this I said, 'I'll have another look for Nude in an hour. I'll take the dog. Perhaps he can ferret her out.'

'Just make sure he stays on the lead,' said my wife, although we both knew there were plenty of worse things for a hen to meet.

*****

When Hoover and I set out on our find-the-chicken walk it took us right round the village, as this was the only way to reach the field behind the house. Progress was slow. Although there were still drifts of snow three or more feet deep – and there had been a couple of days when it was falling so hard we'd been virtually cut off – the surfaces over which people had walked or driven had become hard and glassy. Our road is on a hill, more of a gentle slope really and usually barely noticeable, but now, tottering unsteadily down it, trussed in outdoor clothes and pulled along by a dog keen to find anything worth sniffing, it needed careful negotiation. By the time we'd reached the bottom of the hill, Hoover's quest for canine urine vapour had twice resulted in a sharp tug of the lead and my backside hitting the frozen ground.

The village is built round two roads that form an eye-shaped ellipse. Hoover and I had tottered to where they met, past houses ranging from brightly painted, clapboard wooden bungalows to grander townhouses, set back from the road. The different ages, heights, styles and roof pitches had a pleasantly organic look. Many of their occupants were reluctantly braving the pavements too, as parents walked cold-looking children to the primary school at the opposite end of the village. Others like me, with dogs who still wanted exercising, despite the deep frozen pavements, mingled with them.

Dogs are great at introducing you to people. They force you outdoors and into the company of others, and I'm on nodding terms with a lot of people who first clocked me standing and staring into the middle distance as Hoover took the weight off his digestive tract. I quite like the semi-anonymity of being 'the bloke who walks Hoover', and Hoover certainly appreciates the attention; even on the coldest mornings our travels are punctuated by other people's children asking if they can stroke Hoover or feed him with the bone-shaped dog biscuits I always carry. One of his biggest fans is a little girl with a cheerful, strident manner and a Desperate Dan chin. She rushed up to us so I pulled off my gloves, and fished out a biscuit for her to give the dog.

'There you are, Hoover,' said our six-year-old friend, who then noticed my striped palms and asked, 'Why are your hands funny?'

Since I'm now chronologically an adult, I've come to the conclusion that grown-ups are just wrinkled children with more miles on the clock and more stuff in their heads (much of which they could probably do without); but what is it about adulthood that turns some of us into patronizing, incomprehensible old gits when we speak to small children? This was an affliction I had that morning when I launched into what was intended to be a funny story about sticking my hand to the chicken drinker, eventually becoming aware of the child's blank, 'Why did I ask him that?' look, so I shut up.

'Hoover wants another biscuit,' said my friend, indicating the dog, who was in *'I'm being good'* mode sitting on the freezing pavement, his wagging tail sliding backwards and forwards over the glassy ice. So Hoover got his biscuit, a small child wasn't bored into a coma and we moved on.

Progess was slow. Quite often adults want to join in the Hoover worship, so a combination of this, extended explanations about the state of my hands, and the fact that trying to walk normally on the deep frozen pavements would have resulted in a lot of falling over and ending up with my left boot wedged into my right ear, meant that it took about forty minutes to cover the half-mile circuit to reach the field that backed onto our garden.

Here the snow was deep enough to reach Hoover's chest, and he began lolloping through it. I followed more slowly, wanting to be indoors but aware too that I was lucky to be outside surrounded by muffled, snow-bound beauty. By the time we reached our fence, gobbets of snow were hanging like icy Christmas baubles from Hoover's underside. His beard was similarly decked out, but the dog seemed happy as he stared up at me. From the other side of the fence Svenson heard us and crowed. Good. He was doing his job, protecting his girls.

Despite Jane's misgivings I took off Hoover's lead, then prodded around in the stark brambles that sprawled in front of the fence, but there was no sign of a ginger chicken. Hoover seemed to find all this fascinating, stuffing his face under the brambles, apparently impervious to their thorns.

Ten minutes later and I heard a growl and turned to see the dog tense and excited, backside up in the air, tail vibrating. The rest of him was wedged beneath another bramble bush. Hoover growled again, yelped and wriggled from view. I could hear scuffling and barking and the bush shook, but the tangle of branches was so thick that I couldn't see what was going on. Whatever he'd found, I hoped it wasn't Nude.

Then something erupted from the hole Hoover had made. It was a small rabbit whose impassive face contrasted with its wide eyes. It shot past, running over my boots. Behind it Hoover was struggling to get out of the bramble bush. Not being a great strategist he hadn't followed the rabbit and was getting tangled up, and by the time he was free his prey was shrinking fast as it hurtled down the path we'd made in the snow.

That didn't stop the dog from belting after it, and all I could do was watch him vanish. Perhaps keeping him on the lead would have been a good idea. Still, we were a long way from the road, I knew calling him would be useless as every brain cell and sinew was focused on the chase, and he'd be back once the rabbit had finished outpacing him. If not I would be looking for a missing dog as well as a missing chicken.

I was still searching for one of them when, after an uncomfortably long time, the other finally turned up. I'd been probing yet another Nude-free bramble, and turned to see Hoover standing in the snow about five feet from me. He'd had a great time, and the fact that he'd failed to catch anything – which gratified my townie soul – didn't matter to him because the chase had been enormous fun.

'Good boy,' I said, proffering a slightly damp dog biscuit, and reattaching his lead. There was no point scolding him. The dog had done the right thing by coming back, and would already have forgotten that he shouldn't have run off in the first place.

'We're not going to find any chickens here,' I said, so we resumed our circuit of the village. Soon we were labouring across the snow-covered green, and made our way to a small, half-timbered thirties' building that had once been a tea room, but was now the model railway and newspaper shop (every village should have one). It is run by Veronica and Harry, a couple who could have retired years ago, but like getting up at 4.30 a.m. to sort out the newspapers. Veronica is one of Hoover's very favourite people, because she keeps dog treats under the counter especially for him.

'Have you been a good boy?' she asked Hoover, who was giving her his special 'angelic-but-starving' look as she gave him a biscuit. I decided not to disillusion her with the rabbit-chasing incident. Having got what he wanted, Hoover yanked on his lead and headed for the exit. He always does this, and it causes both Jane and I a small, irrational sense of embarrassment. Fortunately Veronica finds our pet's naked greed endearing.

As he tugged on his lead I handed over money for the local paper, which again involved removing my clumsy outdoor gloves. Veronica stared at my hands and asked if Jane had given me 'six of the best'.

'Something like that.'

Once home, I towelled down the dog, who appeared to be melting into the carpet, and gave him breakfast (something he clearly felt was long overdue, despite already having eaten his own bodyweight in biscuits). As Hoover began his customary culinary demolition, I went back into the garden to hunt for Nude.

The air still had a solid coldness, despite the watery sunshine and hard, cloudless sky. All the hens had congregated next to the fence up which a patch of yellow sunlight was splashed, making this spot fractionally warmer and slightly less uncongenial for them. I noticed Too and Slasher almost propping one another up, each standing on one leg, presumably to reduce contact with the icy ground.

At around sixty feet wide and a hundred feet long, overlooked by mature trees on one side and rows of shrubs on the other, our garden is an enclosed, rectangular space with plenty of places in which a chicken could make herself scarce. In one corner is a wooden summerhouse, across the way from the chicken's shed/aviary arrangement (known to us as 'Beak House').

Kent was once famous for its orchards, most of which have been grubbed up as the trees aged and cheaper, foreign fruit arrived, but our garden still contains a trio of ancient apple trees, and a bigger, equally venerable pear tree. All still produce more fruit than we know what to do with, but had given this up for another year when I trudged past them. Having made yet another circuit of the garden, I began walking back to the house. This involved passing a vegetable patch and its tottering fruit cage, beyond which was a crescent-shaped fence with a gate in the middle that served as a Checkpoint Charlie for chickens. Beyond this was the house and remaining third of the garden, which is reserved for us, although the hens always make a surgical strike on this forbidden territory if the gate isn't shut properly. They know our part of the garden contains a bush where they can hide but we can't get them. Squawks, in particular, is a genius at insinuating herself into it and has spent hours there, presumably waving two feathers at us when we try to get her out.

I knew that Nude wouldn't be loitering in there, because the gate had been properly secured, so I went next door. Joyce (who did laugh when I'd asked to search her garden for a missing chicken) said this was fine by her, but that she wouldn't be joining me.

'I'm going to stay in the warm. If I see anything, I'll bang on the wall or phone you,' she'd said.

Ten minutes later and still henless, I was back in our garden and uncertain what to do next. Eventually I retraced my steps to the rear of the henhouse, where I'd built a nest box. I opened the lid, just in case someone had laid something. A chicken was sitting on the nest – which was a good start – but attempts at moving her resulted in the bird shooting off and vanishing into the garden, with a lot of squawked complaints, leaving behind an eggless straw circle. I had almost reached the house before realizing what I'd seen, and hurriedly retraced my steps.

Nude had yet to go back to the henhouse, and was standing by the half-collapsed fruit cage, muttering to herself at the outrage of being disturbed. She seemed none the worse for her night out, but a little stand-offish with the other hens. Had she seen me and done the equivalent of a pantomime villain ('She's behind you!'), sneaking into Beak House when my back was turned? It was entirely possible. Perhaps she'd had a dust-up the night before and gone off in a chicken huff, or become so engrossed foraging that it had grown dark and she'd been too far from the henhouse to find her way back.

Hens have been domesticated for thousands of years, but still look and sound very similar to their wild relatives, and have retained a lot of their survivalist instincts. They can be very good at blending in with the scenery, and it was highly likely that on my previous garden searches I had genuinely overlooked Nude, which was what Nature had intended. What the hell, she'd survived, and didn't even seem especially hungry. I scooped some grain from the food bin with my hand, but she only pecked at it reluctantly before waddling off like a music hall comedienne hitching up her bloomers.

By dusk she was back in the flock, but as the weary, frozen day sank into another harsh night, I made sure that everyone was safely tucked up in the henhouse before locking them in, this time with my now only slightly uncomfortable hands encased in gloves.

'I had a feeling she'd be all right,' said Jane.

I agreed. 'We were lucky.'

'Yes,' said my wife. 'This time.'

A VISITOR

Death came to the garden almost two years to the day before Nude went walkabout, and it was the reason we'd been so concerned to find her. We didn't want her to suffer the same fate as her sister on that winter's morning.

Everything had started as normal. I'd not long come in from getting the birds out of bed and had let them into the garden, where they'd engaged in the usual, private, food-related things that interested them. Back in the house, I pottered about in my stockinged feet. Jane was in the kitchen, pouring coffee, when I heard her say 'Oh no!' in an anguished whisper. I wondered if she'd spilt something, but then she started shouting: 'Martin! Martin! MARTIN!'

I ran into the kitchen to find my wife transfixed, staring into the garden. I looked outside and saw the frantic flapping of wings, battering uselessly against something lithe and brown: a fox, carrying our cockerel Sven in its mouth. I rushed onto the lawn, yelling as I went. Crashing through the gate, I saw the fox halfway down the garden, loping easily away. At the summerhouse it stopped, looked at me dispassionately then slipped round the back of the building and out of sight. I stood, panting and shaking, then looked round, expecting carnage.

Instead I saw a great deal of movement and heard a torrent of shrieking and clucking, which I'd blotted out as I'd run. Birds were milling about, but everyone still seemed to be there. Then I became aware of a numbness (that rapidly became pain) in the soles of my feet. The ground was hard and the grass white with frost, and I was standing on it in my socks. In my haste, I hadn't stopped to put on some boots.

I shouted for Jane. I wanted to go back into the house and get some boots, but wasn't leaving the birds. It seemed unlikely that the fox would come back, but I wasn't prepared to take the risk. The soles of my feet grew colder and colder as I called, shouting until my voice became ragged.

Eventually the back door cracked open: 'I'm coming, I'm coming,' said my wife. She sounded harassed and unhappy. Boots were dropped on the patio, Jane stepped gingerly into them and walked towards me. I realized that I was still shuddering from the adrenaline rush. We hugged briefly and she said, 'It's all right. Go and get some shoes.'

Having peeled off my icy, mud-caked socks, dried my feet and found a thick jacket, I pulled on some wellington boots and trudged back into the garden. I poked about in the undergrowth where I'd last seen the fox to try and work out where it might have got in. There was nothing obvious, but I realized that the lattice fence panels that separated us from the fields behind gave predators an open view of the garden and its occupants.

'We're going to have to deal with Sven,' said Jane, pointing to something lying in the grass. It was our dead cockerel; on his back, neck twisted sideways, wings a little outstretched, a few downy feathers fluttering nearby.

'Oh no,' I muttered. An involuntary, pointless utterance.

I felt shocked, and surprised that I felt shocked because I'd seen Sven's encounter with the fox, and what had obviously been his death throes. Jane said that it must have been quick, and that was true. By the time I saw the spasmodic flapping of the bird in the fox's mouth, he was probably already dead. As we looked at this pathetic bundle, neither of us moved. I've never been great at handling the practical aspects of death, so it was Jane who dealt with Sven's mortal remains. I clean out the birds, and am an expert in the scraping, collecting and disposing of chicken guano. Jane is not, and sometimes I chafe against this, but she's usually the one who takes charge when the detritus in question used to be a member of our flock.

As she headed back to the house for a bin liner I turned my back on the dead cockerel and began checking the rest of the flock. Everyone looked fine, but then I noticed a small, bloody patch on one of Lewd (the Buff Orpington's) wings. That would need further investigation, but it wouldn't be kind to try chasing her round the garden to have a closer look.

Jane returned with a bin liner and scooped up Sven. He'd gone out in a blaze of glory, doing his job. Although he was old, rheumatic and slow, and stood no chance against the fox, he'd still taken it on. We'd witnessed his final seconds and had scared off the fox before it could go on a killing spree. And that's what would almost certainly have happened.

Hunger and hunting instinct had brought the fox into our garden in daylight, but often these animals seem to go into a killing frenzy, leaving dead or injured pets in their wake, and not taking them for food. We'd been lucky; and for Sven it was the perfect exit.

*****

So I'd found a shovel and dug a deep hole in the rock-hard ground, breaking into a sweat with the effort, then picked up the bin liner and its warm, soft contents. Sven was rolled into the hole, covered with a piece of old cloth and buried. I just hoped that I'd dug deeply enough, and that Hoover, or perhaps another fox, would not have a go at revisiting him.

Our unwelcome visitor meant that none of the birds could remain outdoors. Jane and I decided that for the time being they'd have to be moved into the henhouse. They radiated a mix of nervous energy and a subdued air as we rounded them up and herded them towards it. Even Ulrika and her two teenaged cockerel brothers – who normally put up the most resistance to imprison-ment – followed the others and were soon shut in. After having the run of the garden it would be a pretty miser-able and boring existence, but safer than being left in the open and vulnerable until I'd found some way to make the garden more secure. With everyone behind bars, we headed back to the house, serenaded by the sound of disconsolate squawking.

The fox hadn't been the only uninvited wild guest. Rabbits had broken into our garden too, and we concluded that they had probably helped to attract it. Despite the destruction the rabbits visited on some of our plants and the contents of the fruit cage, we'd found them endearing, especially the babies, who looked as if they'd escaped from the lid of a box of chocolates. They'd also been able to escape from Hoover, whose attempts at bunny chasing had been loud, clumsy and incompetent. Now things had changed. Perhaps the fox would have come anyway, but those rabbits were a source of prey, and another temptation along with the chickens. Our squeamishness about a bunny cull had to be overcome, and they would have to go.

'They dig holes under the fence,' said Jane. 'If a rabbit can get through one of those you can bet a fox could too.'

This left the problem of who we could ask to kill the little buggers off. We didn't know anybody locally with a gun.

'What about David?' asked Jane. She was talking about an old school friend of mine who's a crack shot (and a keen cook). He'd be happy to blow away and then cook Flopsy, Mopsy and Cottontail. 'I think we should invite him down for the weekend,' said my wife. 'On condition that he only gets fed when he's killed something.'

The other issue was how to make the garden fox-proof. My grave-digging experience had shown that putting in ground-level security by burying chicken wire beneath our fencing would be almost impossible until the milder weather arrived, or at least until rain softened the ground.

'We've also got to find a way to fence things off so that the garden doesn't look like a prison camp,' I added gloomily, realizing that whatever we ended up doing was going to involve a lot of hassle.

Life had to go on, so Jane set off for work and I headed for my home office, but found concentrating hard. It was mid-morning when I remembered Lewd's wing and went to the henhouse to find her huddled among the other birds, but I could still see a dried trickle of blood on her feathers. She resisted capture in a spirited way that I found encouraging, but what I discovered next wasn't good. She'd been bitten, probably once, and there was a puncture wound that went through her left wing, which felt warm and puffy. I called the vet, and less than an hour later was walking into the surgery with the patient sitting dejectedly in a cat carrier.

Our vet was a German girl called Anka, who was unusually well versed on the intricacies of bird health. She cleaned up the wound, but remained concerned. 'It may be infected,' she said, reaching for a syringe with which to give Lewd a hefty dose of antibiotics. 'But she is a young bird, and she seems to be very healthy. With luck she will get better.'

Lewd, who'd arrived as one of a pair with her sister Nude, hadn't been with us for long and was little more than a baby. I really wanted her to pull through, but after a few days the flesh of the wing seemed to expand and became a heavy, dead weight. The bird was still eating, but otherwise seemed a little down and listless.

The fox didn't return during the day, but it was almost certain that it had been back to see if there were any more pickings when we weren't looking. Confined to the henhouse aviaries, the hens were becoming restive, and were keen to get back to the garden. Changing their water and filling their feeder had to take place when they were locked up in the henhouse to prevent escape attempts. Otherwise the birds would gather in clumps behind the chicken wire and peer out resentfully like POWs with beaks, as another mealtime rolled round and they still weren't allowed out.

When Lewd's wing started to feel hot I took her back to the vet. Anka was not encouraging.

'It's still infected,' she said. 'We need to really clean the wound properly, and that's going to hurt unless we anaesthetize her.' I asked if anything else could be done if this treatment didn't have the desired effect. 'Well,' said Anka, 'we could partially amputate the wing, but that's a last resort.'

I knew anaesthetizing Lewd wasn't risk-free either, but I left her with Anka, shutting my mind to the likely size of the bill. Lewd was due to be knocked out the following morning, and I spent more time thinking about how she was getting on. When the phone rang and a voice said 'It's the vet' I feared the worst, but was cheered when it said, 'Lewd's fine. You can collect her this afternoon.'

I arrived to find the bird eating greedily and wearing a slightly cartoonish wing bandage. 'Give it a couple of days, then bring her back and we can see how things are going,' said Anka. I assumed that this would also be when I'd have to cough up for Lewd's medical needs.

Examining what had been done without actually removing the bandage, I was pleased that the wing felt cooler, and a little less enormous. Perhaps we'd turned the corner. But the improvement didn't last, and by the time Lewd was back at the vet yet again her wing was hot, and even more bloated than before. We discussed diminishing options. She could be put to sleep, or have the damaged area of wing removed. Taking a deep breath I said, 'We've come this far, we might as well go the whole hog and have the surgery.'

'Well,' said Anka, 'we're running out of time now.'

I was told that the operation would have to take place at a branch of the vet's practice in Rye. A date was fixed, another form signed, and once again I neglected to ask how much all this was going to cost. I didn't want to know.

A few days later the remarkably calm and perky bird sat in the cat carrier on the front seat of my car. I'd kept her apart from the others and she hadn't eaten recently, a pre-op necessity that did not meet with her approval, and she fruitlessly scratched about in her bedding for anything edible. At the vet I stuck my index finger through the carrier bars and tickled the patient under the beak. Having tolerated this for a few seconds, Lewd jerked her head away, turned her back and began more fruitless foraging.

'You'll be all right,' I said, and headed for the exit, reasoning that she now had a track record of surviving being knocked out, and ought to get through the surgery. The phone rang late that afternoon.

'Mr Gurdon?' asked a professionally cheerful voice. 'It's the vet. I'm sorry but . . .'

Lewd hadn't made it. She'd died under the knife. It seemed very unfair that she'd come this far and hadn't survived. After two weeks the fox had claimed a second victim. 'Perhaps it's for the best,' said Jane. 'I did wonder what sort of life she'd have had.'

Soon after that we received two pieces of correspondence, both in buff envelopes. One was a condolence card from the vet, commiserating about 'your sad loss of Lewd'. The other was an invoice for £221.

I paid, pondering the irony that Lewd had only been part of our flock for a few weeks, hadn't had much of a life in chicken terms (no egg laying, broodiness or being the subject of serial bonking), but a single fox bite had made her very nearly as expensive as all the rest of our flock combined.

CAGE FIGHTING

In the weeks that followed the surviving chickens had become ever more fractious and ill-tempered in their secure aviary compound. They were frantic to get out, but the cold snap continued and the ground remained granite hard: too hard for sinking in chicken wire to stop rabbits and foxes burrowing under our fences. In the end I found an old electric heater, installed it in the summerhouse, bought some plastic roll-up fencing of the sort used by road menders to stop people falling into holes, and used this to fence an area between the hen-house and summerhouse.

Once a day I'd let everyone out for some prison-exercise-yard-style walking about, and retired to the only slightly less freezing summerhouse to shiver over a word processor, whilst keeping an eye on the birds. The main problem I had was the still-worsening relationship between the chickens. When Sven had been around, he'd generally sorted out any nastiness. Now his sons were starting to prove a problem. They were growing up fast, changing shape and gaining decorative feathers that definitely shouted 'adult boy'. They were also under-going the chicken equivalent of a teenager's balls dropping, and they'd started sparring. These encounters were still harmless. Neither had much in the way of spurs, but they were doing that rigid little cockerel dance, puffing up their neck feathers and staring hard at one another, then flinging themselves round the run.

These lightweight masculine collisions did not endear them to the girls, and every so often the most senior hens, particularly a veteran Buff Sussex called Peeping Chicken (who was now heading the flock in the absence of Sven) and Ann Summers, would give them a brief battering, but the gaps between hostilities were getting shorter.

'One of them has got to go,' said Jane.

With relations deteriorating I just hoped that the weather would soon ease up enough to make the garden more secure, and give everyone a bit more space. As winter progressed there was still little sign of this and as bird tempers frayed, human ones were suffering too. I was becoming increasingly irritable, freezing my extremities in the flimsy summerhouse, whilst failing to get much work done because of frequent trips outdoors to separate warring birds. When there was peace the summerhouse was so cold that my fingers froze and I kept hitting the wrong keys.

In the end I started operating a chicken rota, with one cockerel and a few hens taking it in turns to wander about in their cordoned-off area of grass, whilst the others stared resentfully from behind the wire of their aviaries.

By now we'd advertised a free cockerel in need of a home, but without much hope of success. Being shirty, sex-mad, egg-free crossbred noise machines didn't make our cockerels an easy giveaway, although they were good-looking birds. Since both were pretty much identical, the idea was to keep whichever one failed to find a new billet.

'I'm not going to give either of them to someone who wants a cheap meal,' I said to Jane after there hadn't been a flicker of interest.

47

'How about Annie and Nick?' she'd asked.

They had been near neighbours in the village where we'd first started keeping hens, and we'd stayed in touch to the extent that when they began keeping hens as well they'd sent an email letting us know. I replied saying that this was brilliant, and by the way, did they want a handsome copper- and gold-coloured cockerel? A week of Internet silence seemed to confirm that they didn't, but one morning my inbox contained welcome news.

'Your cockerel,' asked Nick. 'Do you still have him?'

Our friends had moved from an end-of-terrace cottage to a detached house surrounded by woodland, sufficiently far away from any neighbours that a male chicken giving it some verbal welly wasn't going to create a problem. The method of choosing which bird would stay with us and which would go was completely arbitrary, being dictated by the one that was the least difficult to catch. At this stage neither of our geezer chickens had names, and we didn't feel we'd bonded with them, so letting one go was easy (especially since he was heading for a new home with people we knew and trusted).

However, the bird I finally laid hands on and stuffed into a cat carrier made it plain he didn't feel this was a good idea. He was strong, and tried several times to wriggle free and savage my hand with his beak. Fury was only slightly abated by a ten-mile car journey, and he began bristling again as soon as I removed the carrier from the car. Annie was waiting, and peered inside.

'Oo,' she said. 'He's a pretty boy. Our girls are going to like him.'

When released, the bird made a beeline for the new hens, and instantly began bobbing and weaving round them. The reactions from his future harem to this showing off ranged from mild interest to moderate alarm.

Annie laughed. 'I think they're in for a surprise.'

Back with our flock we soon began calling the remaining cockerel Svenson, and he blossomed without direct competition. That was how things continued until the weather finally became milder. I bought solid fence panels to replace the lattice ones at the bottom of the garden, so that it was completely shielded from view from the outside. I spent hours digging trenches in front of all the fence panels and inserted L-shaped sections of chicken wire, which I stapled to the bottom of the panels. The idea was to create a subterranean barrier to anything that wanted to burrow its way in. Then I added uprights to the top of existing fence posts and slung sections of dark green, plastic roll fencing to these, so that any animal trying to climb into the garden would have to work hard. It wasn't pretty to look at, and there was still no guarantee that a fox couldn't one day find a way in, but it seemed good enough to once again properly release the birds into the garden.

'Well,' said Jane, as we watched them all milling about in a space that allowed them to get away from each other. 'If a fox gets in and finishes them off now, they've had a nice life first.'

EGGSITS

Predators aren't the only reason for departures from the henhouse. Winter is the time of year when you're more likely to find a sad, stiff little bundle of feathers, as someone (usually – but not always – an older bird) has packed up and keeled over. Sometimes it's the apparently strong, active ones that go first, leaving the medically challenged to wheeze on into their dotage.

We once owned a hen called Egghead, who was a serial invalid. A large, lumpy creature with a plume of feathers at her throat which looked like a small beige firework, she had deformed feet, and in late middle age largely gave up walking. Every winter she seemed to have a near-death experience (the most gruesome involving her backside and an infestation of maggots). On several occasions I'd find her sitting next to the henhouse after dark, like an Eskimo pensioner dumped by the family igloo, waiting to die.

We have various cat carriers and a folding dog travel cage for invalid birds, and I'd clear a space in the tool shed and put her in the cage, sometimes with a garden tube heater. If it was really cold I'd find a spot in the corner of my home office and move the caged patient indoors (an act that would give professional chicken keepers hives). Jane was not entirely enthusiastic, either;

coming home from work and inhaling the Egghead-perfumed air she would mutter, 'You've got a bloody chicken in the house, haven't you?' in a tone of semi-mock exasperation. But she has never insisted on chucking out an avian guest.

Tins of sweetcorn or bags of raisins would be bought from our village shop, and these always revived Egghead's appetite. She'd decide that death could wait, although there was usually a tense forty-eight hours as we waited to see whether the Grim Reaper had other ideas. For a long time our ancient, slightly stinky, sit-down chicken would win the day, and after a week (which usually involved some half-hearted attempts to get her to walk a bit more) we'd put her back in with the others. True to type, this death-defying chicken hung on and outlived birds who were apparently far healthier than she was, and was well into her eighth year when she finally faded away.

Winter is also a time when an owner's sin of omission can lead to tragedy. Hens hide feeling unwell for as long as they can, because as flocking creatures the weak are considered liabilities and will be rejected by the others. So when they look a little ropey to us they've often been ailing for days and are in a very bad way. One of my absolute favourite birds, Peeping Chicken (a stately, benign despot, so-called because she arrived in our flock as a teenager who 'peeped' rather than clucked), was well into her dotage and seemed to be coping well with a very harsh cold snap when I found her collapsed in the nest box, beak open, breathing shallowly, using her wings to support herself. Picking her up I found a skinny, skeletal body under a mass of still healthy-looking plumage. The vet was not optimistic, but suggested a course of antibiotics.

Back home I gave the exhausted bird a dose and wished I hadn't. She sank sideways and the life went out of her eyes. My attempts at medication had finished her off. Proof that you can kill with kindness, and that trying to keep her going had more to do with our feelings than her needs.

Subsequently there had been a long period during which our flock seemed to be in rude health, despite the cold. It was after the snows with which Nude had become rather too familiar had melted, the days had begun lengthening and shrubs and trees were tentatively in bud, when I noticed that the self-effacing Brahms wasn't eating. It had been raining a lot, with days of fine, freezing drizzle, and the bird was cold and soaked. I brought her indoors and installed her in the dog cage, which I put on a surface in the corner of my office, from where she watched me warily as I worked. For the next couple of days she barely ate and I feared the worst, but eventually the usual mix of special edible comestibles and a lack of competition got her appetite going again. She quickly grew perkier, in a low-key, shy sort of way.

After a few days of inhaling *odeur de poulet*, Brahms and her hospital cage were moved into the summer-house, which was still a lot warmer than the garden.

Then it was out of the cage and into a chicken run we use for broody hens and their chicks. This reintroduced her to the great outdoors and made her visible to the other birds, but gave her somewhere to roost at night (too long an absence could result in her being forgotten, and treated as a stranger, which would mean getting duffed up). Brahms shared her chicken chalet with Meringue, who resented being imprisoned, and scuttled up and down the wire enclosure like someone who's spent all night on the pavement outside Selfridges and can't wait for the sales to start. Still, at night, she and Brahms hunkered down together for mutual warmth, and a few days later both were back with the other hens.

Winter was almost over when Squawks caught a cold (or a bug known as croup), which resulted in gurgling breath and sneezing. It sounded contagious, and the vet said that it was. So out came the dog cage and in went the chicken.

'Oh God,' said Jane, 'another one!' as a loud avian sneeze came from my office. Later she borrowed my computer to check her emails (this is Jane, not Squawks).

'It's a bit disconcerting,' said my wife. 'She seems to find someone emailing schools absolutely fascinating. You come into the room and she stops scratching about and just stares.'

'I wouldn't take it as a compliment. She's only after food.'

If it wasn't on offer, Squawks had a knack of letting you know that this was very poor by staring at her empty food dish then staring at you in a *'So what are you going to do about it?'* way.

'Shut up,' I found myself muttering during a visit to my malodorous workplace. Even though Squawks wasn't making a sound, the implication of what she wanted was obvious.

I found making business calls accompanied by what sounded very much like a death rattle a bit of an issue, and eventually took my cordless phone into the hall, where only I could hear Squawk's fifty-a-day wheezing. Otherwise she seemed as bright-eyed and greedy as normal, and I had high hopes of a full recovery. Eventually she rejoined the flock in the same staged way that Brahms had done, and it was the end of February when we watched a sneezeless, healthy Squawks, trailing a piece of bramble, lollop after a sleepy fly, which chose not to be eaten by zigzagging out of the way. A daffodil that was in her path didn't have this option and was trampled into the lawn.

'That's the first one I've seen,' said Jane wistfully. 'And it was almost ready to flower.'

'Never mind, they'll soon be plenty of others for Squawks to murder.'

'True.'

Despite the flower battery, Jane was still delighted not to be sharing her home with a sneezing chicken, and once we got back to the house said she'd bought me a present to celebrate.

'Shut your eyes and hold your hands out,' said my wife.

Eyes were closed and something, hard, cold and cylindrical placed in my outstretched hands. A large tin of air freshener.

'Spring cleaning, my love,' said Jane. 'Whenever I go into the office I start to sneeze, so unless you want to move in with the chickens, I'd make use of that now.'

SPRING

SPRING LOADED

March, a month of fecundity and filth, is when spring arrives and all sorts of things start to wake up.

In our garden the filthy element could be found in the waterlogged, muddy patches on the lawn that were the legacies of January and February, and the often foul weather that went with them. The other purveyor of filth was Svenson the cockerel, whose loins had begun warming along with the weather, resulting in ever-more frequent harassment of his reluctant harem.

Lustfulness had never quite deserted him, even in the depths of winter, when the light had been either half-hearted or hard and in short supply. Now that the shortest day had been passed and the nights were shrinking, although they still weren't short enough, Svenson was keen to make the best of what daylight there was – unlike people, chickens generally don't do 'it' in the dark. Night-time is for sleeping. The daylight hours are for eating, laying or being laid, although having said that cockerels – including ours – are happy to start shouting the odds before the sun officially comes up. They want the world to know that other things are rising too.

Of course March can be cold and miserable, but even so, the natural world is still preparing for a summer of love. So in our garden leaves were starting to appear on annuals, and the first of the Spring spring flowers were just beginning to add colour to the slate greys and mud browns of the preceding months. On the farms near where we live the fields were filled with pregnant sheep, so when Jane and I took Hoover out for a walk past the fields where they stomped hormonally about we began looking out for the first, wobbly lambs. If nothing else, Svenson's desire for serial bunk-ups was part of a wider natural order.

In the henhouse (where Meringue in particular frequently hid to avoid being jumped by a cockerel old enough to be her grandson) the occasional egg was being laid – another harbinger of seasonal change. The productive life of a chicken varies enormously, depending on the breed, but I've read that the optimum egg-laying working life of a factory-farm bird can be well under a year. Certainly many are reckoned to be over the hill laying-wise by the time they've reached their first birthday. With the youngest member of our flock at least two years old, and more than one well into her seventh year, we didn't expect miracles.

Still, the intermittent egg production which featured during the winter had started to be replaced by more concerted activity, and we knew it wouldn't be long before Jane and I could lift the lid of the nest box and find a couple of eggs on most days. Varying in size and colour, they would be shades of brown, some off-white, and Slasher has become a regular producer of very attractive green/blue eggs. Her habit of stashing them all over the garden, however, meant that our larder often failed to enjoy the benefit of her labours.

Although I share a house with a woman, a terrier and a splenetic old cat, who are all partial to eggs, or egg-related products, our flock can still produce more than we'll ever need, and we give surplus eggs to friends and neighbours. This had happened recently, after Slasher had spent about a week providing them every day (and actually depositing them in the henhouse nest box), so we had plenty of her distinctive *oeuf oeuvre*. We had given some to Veronica from the newspaper/model railway shop, who had never seen a green hen's egg before.

'You're going to think this is silly,' she said, 'but I cracked one open to see what colour the stuff inside it was. It was just like a normal egg,' she added, sounding almost disappointed.

Another relative of my wife's, who is determinedly urban, refused the offer of our birds' eggs saying, a little bafflingly, 'We do have eggs, but I like to know where they come from.' Waitrose presumably. We decided not to say that we knew *exactly* where our eggs originated, and could often match each one to the bottoms from which they were laid.

On one weekday mid-March morning the owners of those bottoms were still shut in the henhouse and keen to get up. Unlike dark, ice-bound winter mornings when they wanted to remain indoors in a moderately warm huddle, the middle of March is a time when as soon as there's daylight our flock is very keen to get into the world. Before this happens I will go through the ritual of refilling their galvanized drinker with water from a tatty green plastic watering can. Any water left from the day before is chucked onto the lawn. Having something clean to drink is important for everyone's health, although the hens do love to sup from puddles and other fetid places, and it doesn't seem to do them much harm.

With this chore out of the way, I'll turn to the metal dustbin where their food is kept, lift the lid and dump it on the ground, then fumble about inside for the plastic measuring jug with which I will scoop out breakfast. Layers pellets are then poured into the flying-saucer-shaped chicken feeder, hanging on the end of a piece of electric flex in one of the two henhouse aviaries, and into a green plastic dog bowl in the other. It's a hefty thing, and pretty immune to being tipped over by the marauding clientele it now serves.

With everything set for a chicken stampede I will wander over to the summerhouse, which by March is no longer as inviting as a deep freeze, and switch on the radio. Not for our chickens the banalities of commercial radio stations. We go for Radio 4. First thing in the morning the *Today* programme is in full swing, and we think the average passing fox will be scared off by Robert Peston (he certainly scares us). Since 99 per cent of Radio

4's output is people talking, the idea that Jim Naughtie & Co are acting as aural scarecrows is rather an appealing one. As our garden isn't yet overlooked by other houses – and shutting the summerhouse door cuts the noise to a gentle mumble, audible to sharp-eared predators but hopefully not our human neighbours – the half-heard ebb and flow of very English speech radio has become as much a part of the garden as the fruit cage, or Meringue trying to escape from a rampant Svenson.

On this particular morning, Slasher had pushed past one of the trap doors let into the side of the henhouse and into the aviary beyond, and was already getting stuck into the food. The doors of these foot-square entrances are held open by more bits of old domestic flex, hung from screw eyes. Soon everyone else piled out, like gone-to-seed football players emerging from a stadium tunnel. All our hens are pretty stocky, traditional full-sized breeds, chosen because they seem to live longer than ex-battery hens, get ill less and (up to a point at least) behave less aggressively than bantams.

This doesn't prevent them from having other eccentricities. Slasher is a good example. She's an Araucana, and that gives her a built-in streak of perversity, which had manifested itself in erratic egg laying. After that rare week of good behaviour when I'd found something she'd produced almost every day in the official nest box the supply had dried up, and I wanted to find out if she'd actually stopped, or had again decided to deposit her eggs where we couldn't find them.

So after a lot of chasing about I caught the bird, donned rubber gloves and set about exploring her nether regions in a way that was undignified for Slasher and no fun for me. A chicken's bum is known as its 'vent'. There are two bones on either side of this that jut out. When a bird is off lay these get closer together, about a single index finger's width apart. The technique for checking this involves placing said finger between the bones rather in the manner of someone making a 'shushing' gesture, and I've done this many times. When a bird is laying it's possible to get two or even three fingers between the gap, which (gesture-wise) is more 'Oh my God'. Anyway, I was able to give Slasher the double-digit treatment, which indicated that she was laying eggs somewhere, but not in the henhouse. Putting down the by-now apoplectic chicken (who scuttled off and stood at a safe distance muttering to herself until she forgot why she was cross, and went in search of something to eat), I went to look for her secret egg stash, not realizing that I wasn't hunting alone.

The bit of the garden reserved for the birds is accessed by a big wooden gate, which sits in a sort of timber churchyard-porch arrangement designed to stop Mollie the cat and Hoover from making the acquaintance of the birds unsupervised. Latterly Mollie's forays into the wider world had been limited. She found the house perfectly congenial as somewhere to sleep for about twenty-two hours a day, but even she had noticed that spring was cranking into life, and had recently taken to more frequent early-morning outdoor excursions. Evidence of this came in the shape of the spittled clumps of grass she'd regularly been honking up on the kitchen floor, for us to tread on when we came downstairs in the morning to make breakfast and get the birds up.

This particular morning she'd done that again, but I'd managed not to tread in the results. Later, as I began a pre-breakfast search of the garden for Slasher's eggs, I'd assumed Mollie had retired to her basket, so was surprised when I saw something ginger and malevolent slinking towards the fruit cage. Mollie, who hadn't caught anything for years, appeared to be hunting, but how had she got into this part of the garden? It didn't take long to find out. We'd put wire mesh at the bottom of the fence that divided the human and chicken areas, and at one point behind a climbing rose it had been yanked upward and actually chewed through.

'Oh God,' I muttered, 'we've got rabbits again.'

Were they what our prehistoric moggie was hunting? I didn't think she posed much of a threat to the average rabbit, which could easily outpace her, but in her younger days Mollie had become a mass-murdering country sports enthusiast, and had sometimes shown a slightly unhealthy interest in the hens themselves. If she caught Slasher adding to her egg pile, I wondered whether she could – just possibly – give the bird a hard time. Both she and the chicken had vanished in the time it had taken me to find where the fence had been breached, something I wouldn't be able to fix until the cat had gone back to the house. This train of thought was broken by my wife, who was calling from the back door.

'Are you ever coming in for breakfast?'

No contest. I could look for cats, chickens and eggs after I'd eaten.

Back in the house, and halfway through a bowl of cereal, I'd forgotten about both, having retired behind a newspaper, and was harrumphing about the state of the world when Jane screamed. Peering round the paper I saw what appeared to be a mouse jumping through the back-page crossword, its back legs and tail dangling over my muesli. Since the front of the mouse wasn't obviously punching a hole through the other side of the paper – the bit I'd been reading – it took a few seconds to realize what had happened. Our dear cat had done more than regurgitate grass in the kitchen. She'd also caught and bisected a mouse, got bored and dropped its back end on the newspaper, where it had congealed and adhered.

'I know it's going to be funny eventually,' said Jane with a shudder, 'but it's not funny now. Poor creature.'

Folding the newspaper and putting it and the mouse bum in the bin, I decided to forego what remained of my muesli and concentrate on comforting my wife, whose appetite had also waned. After ten minutes doing that I returned to the garden to try and find Slasher, her eggs and a pensionable furry murderess.

I started searching along the back fence where there are shrubs and a small, slightly forlorn fern garden, made my way to the corner where we keep our compost bins, and poked about behind these. Neither revealed any eggs. Then I skirted the boundary fence behind the henhouse, which contained a row of straggly pine trees, and something Jane and I call our dwarf leyland cypress tree. It's a species that's supposed to grow as if it's filled with Viagra rather than sap, constantly hurling itself skywards, creating instant privacy and neighbour disputes in the process. The bloke we bought this one from promised it would go up like a rocket, but years later the thing remains skinny, short and straggly. Peering round the back of this benighted tree I found the ground was egg-free. Continuing along the rest of the fence I circled the remainder of the garden until I was back at the henhouse, none the wiser as to where Slasher had been secreting her egg booty.

The bird herself remained elusive too. Often the rest of the flock move together in a feathery shoal, travelling over the garden in search of seeds, insects, and more gruesome prizes, such as the occasional live frog. When that happens there's a chicken stampede, with frenzied birds yanking the doomed creature from each other's beaks. Even then, Slasher would often be on the periphery, doing her own thing, which sometimes involved high-speed running about. She is, by a considerable margin, the fastest chicken we've owned, and will hurtle from one end of the garden to the other as if the Four Horsemen of the Apocalypse were on her tail.

Not this morning. She'd really gone to ground, and I was just about to postpone my search when there was a scrabbling noise and I turned to see something brown shoot out from under one of the dwarf leylands and bolt across the garden to the opposite fence. It was neither Slasher nor Mollie, but a very large rabbit. I watched as it flung itself under the breach I'd found in the fence panel, which shook as the animal wriggled its way through and vanished.

'Bugger,' I thought. 'We really must find someone with a gun.'

Then there was a second commotion, perhaps initiated by surprise caused by the first one. This time it came from our semi-collapsed fruit cage, a rectangle of raised vegetable beds and fruit trees more or less covered with netting. I could hear furious clucking, then something ginger and furry wriggled into view, followed shortly afterwards by a very angry chicken. Mollie had found Slasher and, far from being cowed by the encounter, Slasher had lost her rag.

Moving remarkably quickly for one so old, Mollie made for the rabbit's escape route as an enraged chicken rapidly gained on her, but this wasn't the cat's only problem. Svenson was now bearing down on her as well, flapping his wings to aid momentum, and Mollie became the object of an avian pincer movement. They would intersect where the rabbit had made its escape. Only one of them could actually fit through the gap in the fence so Mollie vanished behind the rose bush, leaving Slasher and Svenson to swear impotently at her. As the cat flap into the house clicked shut, the other birds joined in the shouting and the garden soon echoed to the sound of hysterical shrieking.

I made my way to the fruit cage, and looked at the spot where the cat and chicken had emerged. In the long grass at the base of a straggling vine I could just make out a clutch of green eggs. There were at least thirty. Some looked fresh and recent, others had clearly been there for weeks, and had bleached, dirty shells. Thanks to Mollie, I'd finally found Slasher's stash, but dealing with it would have to wait. Once again my wife was calling from the house.

'Come on love,' she said. 'If we don't go now we're going to be late.'

As she said this, yet another commotion broke out. I turned to see Slasher bolting down the garden, Svenson now close on her tail. The cat might have escaped his attentions, but he was determined Slasher wouldn't, and what he wanted from her was painfully obvious. Usually a very successful escapologist, Slasher made the mistake of scuttling towards the compost bins, and a dead end. As I turned back to the house, the disgruntled sound effects indicated that there had been an avian encounter next to the romantic plastic vats filled with rotting teabags and peelings, and that the pleasure had been all Svenson's.

DOVE STORY

Spring is a good time for acquiring new birds (because looking after them is generally less uncongenial at this time of the year), and that morning Jane and I were setting off in the car to expand our flock – but not with more chickens. Although they didn't know it, our hens were about to share their personal space with six doves.

My beloved is a serial hinter, and she'd been keen to try keeping doves for ages. Before that there had been the beekeeping idea. I was less keen on bees for reasons of not wishing to be stung, but Jane did the research and even enrolled on some beekeeping courses, but eventually began having misgivings too. Her mother had reacted badly to insect stings, and increasingly Jane was displaying similar symptoms.

'And you'll be taking them to the vet in matchboxes,' said our London friend Jennie, who shares her home with an irritable cat called Polly, views our menagerie as rather peculiar and – in the case of the dog, whose one visit to her house culminated with him piddling on her duvet – actively unsavoury. The other thing that put us off bees was a missive from the British Beekeepers Association about pulling the heads off selected members of your hive and sending them to a lab to check for signs of disease. Bee numbers have declined alarmingly in recent years, so this is both understandable and important, but the thought of bee beheading made us shudder, and for the moment at least, bees were off the agenda.

On the other hand doves were low maintenance, would not bring Jane out in welts and did not require decapitation. They would make an attractive contrast with our hens too. We really couldn't see what there was not to like about them.

'You don't want doves,' said my Dad. 'They'll crap all over the front of your house.'

With these cheerful words ringing in my ears I bought a dovecote (a posh term for a dovehouse) from a bloke in the charming market town of Tenterden, who made them in his garage and stuck them in the front garden of his bungalow next to a 'For Sale' sign. They looked as solid as anything sold by birdhouse specialists and were a lot cheaper. The model I selected was hexagonal, had two floors of one-room dove penthouses, and sat on a large pole. I chose it because Jane's serial hinting hadn't stopped, her birthday was looming, and I couldn't think of a better idea for a present. If its occupants decided to visit our house and make their own entertainment by crapping all over it, at least Jane would have got what she'd wanted first.

Once the dovecote was standing at the far end of the garden the hens gave it a quick once-over, decided – wrongly, as it turned out – that this wasn't going to be a harbinger of food, so wandered off and ignored it.

'That looks really nice,' said Jane. 'Now all we need are some doves.'

Which was why, after the excitement with Slasher, she'd called me back into the house. We'd spoken to a nice lady in Essex who bred doves and was a fund of useful advice, but didn't have any birds for sale. Then we found a bloke selling them on the south coast who sounded more than a little brusque on the phone but, we thought, never mind. We weren't inviting him into our lives, only his doves.

As we drove to meet him that morning, I told Jane about Mollie's encounter with Slasher and Svenson, the subsequent chicken bunk-up behind the compost bins, and the discovery of Slasher's illicit egg pile.

'Some of them look ancient, but I can't really tell which are the new ones, so I think we'll have to chuck them all away. It's such a waste,' I said as I steered the car into the dove breeder's drive. He lived in a forbidding farmhouse faced with grey render and surrounded by tired-looking buildings, dead tractors and other indeterminate bits of moribund machinery. The dove breeder had painted some unfriendly signs on his tatty drive to deter uninvited visitors from turning or reversing on it, and although the place was on a busy road, it had a slightly alarming backwoods feel. Had we heard someone utter the words 'Squeal like a piggy, boy,' in a Deep South American accent it wouldn't have seemed entirely out of place.

The Gothic-horror-like ambience was heightened by eerie flapping and banging sounds. Peering round the outbuildings we found these were being made by a magpie in a large cage, which sat in the middle of a field adjacent to the farm.

'Yes?' It was a voice that brooked no questions, or bothered with trying to sound pleased to see us, and belonged to a large man, aged about sixty, wearing camouflage and the wrong kind of beard; a large grey, sub-Catweazle affair surrounding a mouth that didn't look as if it had cracked a smile in thirty years.

'We've come about the doves.'

'You the people who phoned?' Had the voice softened slightly? Perhaps.

We said yes.

'This way.' The man ambled like a green-clad yeti towards one of the outbuildings. Inside what had once been some sort of animal shed there was a strong but not unpleasant smell of birds, and doves of various ages and breeds. We chose four youthful, pristine-white birds – two males and two females. Showing unexpected gentleness, the camouflage man showed us each bird in turn, then paired them into a couple of cardboard boxes. He provided us with detailed DIY leaflets on how to look after them (which he more or less read out in their entirety), sold us some rather expensive food, jars he called 'ramekins' to put it in and relieved us of something in the region of £40.

On the way out I saw the magpie flapping and banging against the roof of its cage. The man followed my gaze: 'It attracts other magpies, and foxes too. Means I can shoot 'em.'

'Oh,' we said. 'That's nice.'

Once home we'd been told that the doves would need to be confined to the dovecote for about three weeks, and to make sure of this I'd covered it in a fine black net, so that it looked a bit like the head of an Edwardian lady motorist, swathed against the dust. The doves, who'd seemed relaxed with the camouflaged man, found me less congenial. Getting them under the mesh and into their accommodation was difficult, particularly since it necessitated clambering up a collapsible ladder, which wobbled alarmingly on the uneven ground or became unbalanced as it sank into it, and by the time they were installed everyone was feeling the strain.

I hadn't banked on the twice-daily feeds involving climbing up and down the ladder every morning and evening, clutching little bags of food and a small watering can (normally used to keep our house plants alive) to refill the birds' water containers. The mesh, which was partially stapled to the wooden dovecote, draped down below it and was secured round the post with an elastic tie. Removing this caused it to flap about and me to worry that we might have escape attempts at mealtimes.

The doves were far more interested in cowering in the corner of their little rooms and waiting for me to go away. Dovecotes have entrances called pop holes, often with small ledges in front of them, where the birds can perch, and where I put the small clay ramekins of food and water. The doves frequently knocked these over, and too often I'd untie the bottom of the mesh and watch as one fell into the long grass, where it would take forever to find.

I felt I had developed definite relationships with the other creatures in our household, even if these were of the *'Now you've fed us you can go away'* variety, but as the weeks ticked by, the doves seemed to find me more alarming rather then less. My arrival at the top of the ladder resulted in agitated cooing noises and a lot of crashing and banging about as they shrank to the furthest reaches of the dovecote. One morning I found a dove that had got itself partially tangled up in the netting. Feeling its heart pounding against its little chest I thought it would die of fright before I finally detached it and posted it back into its birdy B&B.

When it rained I would slip on the ladder. When it didn't I would come back to the house and find dove poo under my fingernails. My wife remained charmed by the doves and amused at the way I was dealing with them, but I had a vague feeling that things were not going to plan.

When the time came to release them my father and stepmother Jenny came to stay and decided to record the happy event with their cameras. I wobbled to the top of that bloody ladder, which I'd grown to hate, clutching a small screwdriver to prise off the staples that partially held the plastic mesh in place, a process that caused much terrified cooing from inside the dovecote. Eventually I was able to remove the mesh completely.

Jane passed up some food, which I sprinkled on the dovecote ledges as an incentive for its occupants to emerge, and retired to a safe distance and waited.

After a while a white head poked out of one of the pop holes, and a small round eye blinked down at us. Eventually, the eye's owner tentatively came out, stood on the ledge, wobbled, fell off, then flew unsteadily to the eves of a neighbouring house, where it roosted, and doubtless crapped. Soon it was joined by the others. They all steadfastly ignored the food we proffered, and showed no interest at all in returning to the dovecote. When it became clear that the doves had no immediate plans to come back to our garden we trooped into the house for a cup of tea.

As it grew dark I went outside again. There was still no sign of the doves outside, so I tentatively climbed the ladder and found their dovecote was empty too.

This wasn't quite the last time we saw them. They'd appear on the neighbouring house and occasionally fly over our garden, but they never returned to the dove-cote, and appeared to be around less and less. Then one morning I found a small bundle of white feathers. The camouflaged man had used an ink stamp with his name and phone number, which was applied to the underside of each bird's wing feathers, and as I swept up the pathetic remains I noticed some of them had traces of this. The rest of the bird had become dinner for a passing hawk. We never saw the remaining doves after that. Perhaps they'd been scared away to the woods at the back of the village, but I did wonder if, just possibly, instinct told them how to find the farmhouse we'd bought them from, and that they'd simply gone home.

I didn't fancy following them there and buying some more. In fact, I wasn't convinced generally by dove ownership, which seemed to involve ladder-related personal jeopardy to feed pigeons in camp suits who wanted to be somewhere else. I kept these thoughts to myself, until my wife made a wistful remark about finding another breeder and trying again.

'I don't suppose you want to climb up and down that sodding ladder twice a day?'

'No, but if I'm at home, I'll hold it for you.'

APRIL SHOWER

So April began pretty much as March had – doveless – and the only evidence that the dovecote had ever been lived in was that it was covered in guano and contained one sad little egg, something that reminded me of the eggs Slasher had deposited in the fruit cage. With all the excitement I'd quite forgotten to remove them, so I decided to do that as a means of putting off cleaning out the dovecote.

Clasping an old carrier bag I made my way into the fruit cage's jungle-like interior, followed by Hoover. The dog sniffed the pile of eggs, which had grown bigger since I'd last seen it. I squatted beside him and began carefully picking them up and depositing them in the bag. Hoover found this deeply interesting and, tail wagging, kept shoving his nose into the egg pile, then into my face, causing me to almost overbalance.

'This isn't being helpful.'

I didn't want to fall over and put my hand into the eggs, as some of them had been there for a very long time. In the end I clasped Hoover's collar and hauled him out of the way, but this only encouraged him to come back when I let go.

'Please Hoover,' I hissed. 'Just go and do something else.'

As I spoke the dog suddenly became rigid with excitement. The hair on his back bristled and he began squeaking. Then he charged. I lunged at his collar but it was too late. Hoover had seen a rabbit. He crashed through the pile of eggs and out of the fruit cage in pursuit of a fast-receding rabbit, scattering chickens in his fruitless chase. His quarry had a good head start and had flung itself under yet another hole in our fence long before our dog could get anywhere near it.

Even from a distance, I could see something black and viscous smeared up Hoover's legs. As I surveyed the shattered eggs and inhaled the rancid smell left by some of the older ones he began licking this with relish. I felt slightly sick. Suddenly the idea of climbing a ladder and chipping fossilized bird crap from inside the dovecote took on a new appeal, but before I could get on with that I'd need to get a bucket of water and some gloves to clean up the fetid mess left by our dog. I had planned to leave a couple of eggs in the fruit cage, marked with pencil crosses, to see if Slasher would keep laying new ones on the same spot, but now I doubted she'd want to go anywhere near it.

As I began work on the dovecote, a pick-up truck pulled onto our drive, and a group of builders emerged and clattered into the back garden. We were having a two-storey extension built at the back of the house, thanks to Jane's expansionist tendencies, and they would need tea. The conversation between Jane and I, which preceded their arrival, had been fairly typical of one of the ways that our marriage works.

'We could do with the space,' she'd said.

'But there are only two of us.'

'And all our stuff.'

'We could get rid of some of it.'

'And the dog, and the cat.'

'They've got enough room already.'

'I just think it will suit the way we live better.'

So we put in for planning permission, using an architect who appeared to be a mild-mannered, ironic sort, until he met the extremely youthful planning officer who'd decide on our application. The pair hit it off in the way that railway locomotives do when they collide head-on. One clearly felt the other should be on the teat rather than approving his plans, the other made it very plain that she knew who was in charge. The dislike was palpable in their body language and written exchanges, and when our extension was rejected we weren't in the least surprised.

'Result,' I thought. 'No disruption, mess and half living out of cardboard boxes for six months, or trying to work in the middle of a brick-dust storm.'

'We'll get the plans redrawn,' said Jane, reaching for the phone to call the architect, who needed several minutes' soothing before he could be persuaded to talk about what we could do to keep his planning office *bête noire* happy. Plans were redrawn and resubmitted and the sound of egos banging into each other resounded over our part of Kent for several months before something that looked very much like a smaller version of our existing house was given the official nod. A third of our possessions were packed away, and I prepared to live and work for the rest of the year under a state of siege.

So that's how I found myself making tea for a mostly cheerful but slightly nervous quartet of men in overalls. Builders have a reputation for being a pain, but that's a two-way street; the good ones could dine out on stories about Gorgon-like clients (as long as they didn't mind not working very much afterwards). With the exception of the man who was running the job – whom we'd used before and liked very much – none of us had met previously, and the idea of having the client in the house most of the time when they were working probably didn't fill them with much enthusiasm.

'Sorry, but there's going to be a bit of mess today,' said the Boss Builder, a stocky man in his mid-forties with a fuzz of closely cropped hair, mournful eyes and the profile of an Easter Island statue. If you could give a

Leonard Cohen song facial features, they'd look like this man, who in fact has a ready laugh and people skills that as a journalist I wish I had, because they'd open a lot of doors.

'Hello Hoover,' said the Boss Builder as our dog jumped up at him. The pair had met before and had made friends. 'What's happened to his legs?'

We looked at the fur on Hoover's legs, which was matted and a little off-colour. I tried to explain, something that seemed difficult to do succinctly. Nobody laughed (although it was obvious that they wanted to), and nobody mentioned the mouldy odour, but once I'd brewed up Hoover and I headed upstairs to the bathroom so that his soiled bits could be cleaned. A friend once described Hoover's character as 'essence of dog', but he's all terrier, and his genetic make-up does not contain a cell of water spaniel. Bathing was a process he loathed, and he spent the next ten minutes cringing and shaking in the bathtub as I sluiced the revolting egg gunge from his paws and legs.

Outside hammering and banging indicated that work had started. Having finished cleaning our egged-up dog, I lifted him from the bath and he bolted for freedom, tail thrashing as he scuttling downstairs to celebrate his escape by running up and down the hall, pressing one side of his face against the carpet as he did so. This ritual completed he made for the back door to eyeball the builders, and I followed, opening it to see what they were doing. They were demolishing the conservatory.

'It won't take long to get this down, then we'll get the floor and the patio up,' said a rounded, smiley, thirty-something man I came to think of as Number One Builder, while one of his colleagues nodded. This builder was in his early twenties and wore a pair of designer glasses. He didn't fit the bum-cleavage/roll-up-fag template at all, and was a bit of a male Bimbo. This was not something that could be said of the final member of this foursome, introduced to me only as 'Knocker'. Knocker didn't go in for conversation. Of indeterminate middle age, he had a bullish build, a broad, florid face that indicated a lot of hard living, and a stillness that under other circumstances might have been menacing. The name intrigued me, so as they worked I asked the Boss Builder how Knocker got to be called Knocker. He laughed.

'Can't you guess?'

At this point Number One Builder interjected and said that he liked our chickens. 'They're good-looking birds, but do you want them in your flower beds?'

I looked up to see the gate wide open and the entire flock keenly grubbing up a flower bed Jane had only recently weeded and planted up. Chickens might not be very bright, but they have something approaching a Colditz inmate's skill at finding ways to escape from places where their owners want them to be, and get to others where they don't. That this had happened now was my fault, but was still irritating.

'Excuse me,' I said, and went out to evict them.

The birds had done a good job of eviscerating plants and churning up the bed and its contents, and were not keen on being moved. I spent the next ten minutes yelling, flapping my hands, and trying to chivvy them out. I'd succeed in getting some of them to go through the gate then turned my attention to the others, only for the first lot to wander back in. Knowing that the builders would be watching this pantomime I made a point of not looking at the house or blaspheming at the birds.

In the end I got a handful of grain and chucked it theatrically through the open gate. This caused Svenson to make his special *'The food's this way'* clucking noises and run after it, with almost everyone else in tow. I followed, and walked briskly to the far end of the garden, followed by a determined posse of chickens, who clearly thought I'd feed them all over again. They were half-right. I made for their food bin, extracted another handful of layers pellets and chucked them towards the very bottom of the garden. These were pursued by everyone, and I beat a hasty retreat back to the house, not looking behind me as I went, banged the gate shut, then turned to see something black and feathery shoot by and hurl itself under a large bush, which I didn't have a hope of penetrating.

'Oh God, Squawks!'

I'd been shadowed and outwitted by our least inconspicuous chicken, who was now back in the wrong bit of garden, wedged in human-proof foliage. I knew it could take hours before she'd come out. I looked back at the house to see three builders watching this with frank, amused interest, except Knocker, whose face was unreadable, but he was watching too. Shrugging self-consciously, I trudged indoors.

Retiring to my home office I was soon aware of a rhythmic, stop/start vibration coming up through the floor. It felt like tiny earth tremors and made concentration difficult, which was boring because I had an article to write. My hands touched the keyboard, but the screen remained blank. The floor continued to judder and vibrate. Eventually I wandered to an upstairs window overlooking what was already turning into a building site and stared vacantly at the activity below. Bits of conservatory were vanishing: white-framed glass rectangles concealing three builders walking to and from the pick-up truck, a bit like a small army of ants, piling up the remains of the conservatory then returning to dismantle some more. The exception was Knocker, who was using a giant mallet to give the patio his special attention. Swinging the mallet in a brutal arc, he casually got on with destroying the spot where Jane and I used to spend summer evenings. This also explained the little earthquakes.

I looked beyond him to the now-shut gate which kept out the hens, and realized that most of them were clustered on the other side of it, standing as mesmerized as I was by Knocker doing what obviously came naturally. Only Squawks remained incognito in her bush. An hour later, when I gave in to displacement activity and went downstairs to make everyone another cup of tea, she still hadn't emerged.

GIRL TROUBLE

If the garden was a hive of human activity, its other occupants, and those in the surrounding countryside, were keeping busy too. As Squawks continued to evade capture, the first daffodils were flowering and the fields were alive with the sound of bleating lambs. Svenson was more than up for 'it' and Meringue and Brahms had good reasons not to be pleased. They weren't the same reasons, but the results were pretty similar. Svenson finds Meringue irresistible, but Brahms is a turn-off. This has led to lives spent on the run, and periods of hiding out to avoid being chased off, or chased after, as these events were getting more and more frequent. During the winter Svenson's testosterone hadn't exactly hibernated, but it had not squirted itself through his system with quite the same fervour that started to become a feature from late March onwards, and really kicked in when spring got properly underway.

In colder climes he was still keen on a bunk-up, particularly first thing in the morning, but mostly, when it's raining incessantly, hail is battering the garden and turning the lawn into mush, or everything (bar some hiding places under the trees) is covered with snow, he goes through longer ardour-free periods. This provides some respite for his girls, and both Brahms and Meringue make the most of not being constantly harassed.

Whatever the season first thing in the morning is still the danger zone sex-wise. After a night in the coop, Svenson's loins are aching, and he seems to enjoy coupling alfresco. This results in behaviour that is sometimes devious or unattractive, and on occasions frankly disgusting.

When I open the henhouse doors Svenson is never the first to emerge. He will usually appear after two or three of his more assertive conquests have thundered out (Ulrika, Bella and Nude are usually quick off the draw). Then he will make an entrance, running into the throng like a flash Channel 5 game-show host. At this point he might have a good crow and target one of the hens, dropping a wing so that its feathers fan out towards the ground, and doing a little sort of sideways run/ pirouette. This looks pretty stylish, but mostly does not impress the apples of his eye who will ignore it, especially if there's food about, or something interesting (and preferably alive) to peck at in the grass.

Svenson might have several goes at his boy-band shuffle before giving up and lunging at someone, who will generally run away. Frequently in the middle of this ritual, he will remember that he's rather constipated. Chickens will poo in their runs and houses, a lot, but often make an effort to dispose of their doings outside. This is especially true of broody hens, who spend hours sitting on their nests, only to express themselves fervently when removed, ostensibly to be fed. It's both a weight off their bowels and a means of keeping the nest area a little cleaner.

So, if Svenson has failed to put his masculine surprise package into breakfast-time action, he will often have another urgent delivery. Any pre-teen schoolboy watching what happens next would be convulsed with lavatorial hilarity. Svenson will stop, bend his knees to squat slightly, and make a long, high-pitched wheezing noise.

Jane and I once holidayed in the Far East, and noticed that a member of our party adopted an identical pose every time he took a photograph. It reminded us of a sloth relieving itself, and was further enhanced by matching facial contortions. This man was fabulously English in his ability to spot prosaic details in exotic locations. We were all in a funicular rail car, making its way up a mountain overlooking Penang, with a fantastic view of the city below. Pointing to the base of an electric fan in the car's roof he said: 'Ooh look, a circlip,' As he squatted and screwed up his face and took another photograph, his wife remained unimpressed. 'I don't like heights,' she confided, 'that's why I'm terrified of vernaculars.'

Unlike her husband, Svenson won't actually narrow his eyes with concentration and effort, but anyone watching might imagine him doing so. After completing this fundamental task he will leave a steaming, mini-fondue-sized evacuation on the grass, which I must then endeavour not to tread in. Pleased and apparently energized by this, Svenson will then get back on the case, running after his girlfriends, who will probably have taken the opportunity to sod off to the other side of the garden.

Sven, Svenson's Dad, was less of a sex pest, and in later years rheumatism rather dampened his enthusiasm. He became something of a sit-down cockerel, spending extended periods with his legs tucked in under his wings, keel on the lawn, pecking contentedly at the grass. Then he would remember that he was still a real man, lever himself to his feet, and beat a stiff-limbed path to the nearest potential girlfriend, who would see him coming, and generally run off. This was frustrating for Sven, because he found it difficult to keep up and so would often retire unsatisfied, but at least this saved his love interests from an often-uncomfortable coupling.

They have no such luck with Svenson, who can spend hours hurtling from one end of the garden to the other in pursuit of avian totty. He's fast too, moving like a professional (if feathery) rugby player. Anyone who does not see him coming is likely to be enthusiastically tackled and mashed into the ground. Even if they do, and start running, quite often he has momentum on his side. For Meringue in particular, this can lead to repeated, unwanted coming togethers, when all she wants to do is drag worms out of the ground and mind her own business.

Sometimes an element of strategy will creep in, and Svenson won't always make his real intentions plain. So instead of attempting instant rape and pillage, he will make a less brutal entrance, or stop close to a knot of girlfriends and start to peck at the ground in a rather showy way. This will be accompanied by almost staccato clucking, his *'The food's over here, baby!'* noise, and it usually has the desired effect of making a gaggle of chickens rush to his side to see what's on offer.

Sometimes there really are edible treats, to which his favourites will be given first pecking rights. Sometimes there won't be, but he will make such a song and dance, head bobbing up and down as he keeps up a constant 'cluck-cluck-clucking', that they will all peck and stare hard at one piece of ground to see what it is that's getting him so excited.

The answer will soon become obvious. Since they will be concentrating on the imaginary food, and not Svenson, with their heads down and fluffy backsides pointing temptingly skyward, he'll get his wicked way. Generally, all his girls get jumped from behind, and as the nights become shorter and the days warmer, so the regularity of Svenson having the hots increases, which means that Jane and I – whom he treats with mistrust at the best of times – find ourselves regarded as possible love rivals.

Unlike some of his cockerel predecessors (including a bird called Zorro who had a nifty line in assaulting wellington boots by taking a run and then jumping up and down on them) Svenson isn't actively aggressive. He will confine himself to some macho wing flapping when he sees us coming. Occasionally, in moments of extreme duress, when I've scooped up one of his girls to examine her (or, in the case of Squawks, detached large bits of undergrowth which she has been carting round), he will make agitated clucking noises whilst engaging in karate-style flying leaps at groin level.

These would probably hurt if we ever connected, but Svenson has decided on discretion, and does his kung fu displays about ten feet away from where I'm fiddling with one of his ladies. It's a bit Bruce Lee, a bit balletic, and utterly useless. He means it, but he doesn't mean it that much, something he demonstrated later that afternoon after the builders had packed up for the day, and Squawks had emerged from her bush to continue vandalizing the flower bed, where I'd finally caught her. I'd carried the struggling bird back into 'her' part of the garden, and Svenson reacted to this by flinging himself uselessly at nothing in particular, until I released a highly annoyed Squawks – who was soon highly alarmed as Svenson made it obvious *exactly* what it was that he'd missed about her.

The following Saturday morning, as Jane and Hoover went for an extended weekend walk and I'd finished a happy half hour cleaning out the henhouse, Svenson was still a bubbling vat of testosterone. As excessive leaping about took place I trudged up the garden to what was left of our house, with the idea of getting out of my chicken-poo-enhanced clothes and having a cup of tea.

Bidding the cockerel a two-fingered farewell I negotiated the trenched war zone that had been the site of our conservatory, still serenaded by angry *'Oi! You come back here!'* crowing as I tottered along the bowing planks that had been laid over what would become the extension's foundations, managing not to fall in or collide with the digger and other builders' tools.

Every time I did this I enjoyed a small sense of triumph, and once back indoors still had that warm feeling of personal well-being, even after discovering that one of my socks was caked in something foul. Pulling on a clean pair, something in my lower back went 'twang!' and I rolled backwards onto the bed, feet pointing skywards, as muscles I normally didn't notice went into spasm. When Jane and Hoover returned I was strapped into the tired, elastic surgical truss we keep for back problems, muttering about how much Velcro chafes against flab and walking like a ramrod.

'Never mind,' said Jane. 'Take a painkiller. I've got some news. You know that little livestock farm outside the village?'

I nodded and winced.

'Well, we walked past it and I got talking to the farmer, and he's got a colony of wild doves in his barn. He'd love to get rid of some of them, and we can have as many as we like for free. He says he'll call when he's caught a few.'

'Oh,' I said, 'great.'

BOY TROUBLE

As spring matured we experienced both an absence and an unexpected arrival. Having semi-skated over the muddy ellipse where our conservatory and patio used to stand I was accosted cheerfully by the Boss Builder.

'Just so you know,' he said. 'Knocker won't be in today, and maybe not tomorrow either.'

'Is he all right?' The man's dark-red complexion did make me wonder. Although he was one of the toughest-looking human beings I'd ever met, I could imagine him being carted off to hospital on a gurney, attached to a lot of wires and alarming equipment that beeped.

'All right?' Boss Builder pondered the question, and his usually serious features broke into a grin. 'Yeah, he's fine, but he was fifty at the weekend and he's been on the lash. He'll surface when he's feeling better.'

Having been told about one piece of male bravado, I went into the garden to deal with another, because although the promised doves had yet to materialize we had, reluctantly, become the custodians of Barry the rheumatic cockerel, whose presence was driving Svenson berserk with jealousy. Jane and I are easy prey for an avian sob story, and Barry wasn't our first chicken waif. Beryl was a scruffy little brown hen found apparently lifeless by an electrician working under a portable building. Deciding that he didn't want to share a confined space with a dead body lying in a puddle, the electrician reluctantly got hold of the corpse to remove it, and was surprised when it twitched. He phoned friends who rescued and nursed wild birds, took the near-dead chicken to them in a hessian sack, and with the aid of a hair dryer they revived it. Having nowhere to keep the bird, they asked their vet to find someone who might take it on. The vet phoned Jane and I, and Beryl moved in with us.

She had long shuffled off this mortal coil when we had a telephone call from a chicken-keeping acquaintance who was moving house and could not take the birds. The move was proving particularly fraught. Homes had been found for the girls, but nobody wanted Barry, a Marans cockerel, which meant he was large and heavily built, had the gait of a Regency beau, and was clad in speckled grey plumage.

'If I can't find someone to take him on, he'll have to be put down,' said Barry's owner, who was clearly in some distress. I found myself saying that of course we would take him in and find him a permanent home. Jane was out, and I wasn't sure how she'd react, but there was no going back. 'It's a bit late to say no, isn't it?' was her only reaction, when I broke the news.

So it was officially decided that Barry would come to us, but could not be put in with our flock. Introducing a second male bird was impossible. Barry and Svenson would square up with bloody consequences, so a way had to be found to keep everyone apart. Normally, the hens have the run of our garden, but when we're out we confine them to the home-made aviaries attached to the garden shed where they roost. We decided that they'd be corralled there for the duration of Barry's stay, which we hoped would be short.

I fenced off an area behind the summerhouse, and installed a temporary henhouse. Although it resembled a big, windowless maximum-security prison for dolls, it was actually designed as a rabbit hutch, and had the feeding and watering paraphernalia needed by chickens. Barry seemed to be entirely at home. Some cockerels are crazed psychopaths, intent on battering anything they might regard as a rival (including people), but despite his imposing appearance, this one turned out to be a sweetheart. He wouldn't quite eat out of our hands, but he was used to being around people and had been very well looked after. When feeding time came I never had the impression that he would try taking a chunk out of me with his spurs.

If human beings weren't an issue, Svenson certainly was. Although the pair couldn't see each other, hearing wasn't a problem, and the presence of another male bird in his garden was clearly an outrage for our cockerel. Imagine two drunken blokes at opposite ends of a pub bar, trading insults but never actually coming to blows. This, in chicken terms, is what we had, but with a lot more volume.

So the sound of mechanical digging was interspersed with an endless stream of 'cock-a-doodle-do' bellowing, which roughly seemed to translate into:

'Do you want some then?'
'Yeah. What are you going to do about it?'
'I don't have to do anything about it.'
'Come here if you think you're so hard.'
'No. You *come here.'*

Fortunately our garden is not overlooked by other houses, but even so we knew this couldn't go on. It would only be a matter of time before someone complained, and it would be difficult to argue that they didn't have a point. One of our near neighbours, who has a house filled with teenaged boys and earns his living welding railway tracks, works shifts and has never complained about the noise from our birds. The idea of disturbing this man's precious sleep was something that gave us the willies. Help, fortunately, came in the form of a neighbour's drinks party.

Clasping a canapé in one hand and a glass of red wine in the other, I was steered by the hostess towards a prosperous-looking couple in their thirties, told they'd just moved to the country from somewhere like Clapham, and 'they've started keeping chickens'.

'We love them,' said the quietly glamorous woman, as her husband, whom I gathered did something in the City – which just possibly means he doesn't any more – nodded in agreement.

'And our daughters love them too,' she added brightly.

At this point inspiration struck and I found myself manoeuvring them into a corner and saying something along the lines of, 'What you need to complete your rural idyll is a cockerel called Barry.'

'We do?' said the man. I couldn't tell if he meant it, or whether he was being polite to another guest at a party where he didn't really know anyone, and actually wanted me to get knotted.

'Oh yes,' I said, pressing on in a way that I hoped wouldn't sound desperate. 'He's a lovely bird, very gentle. If you take him on I'll give him his medication, and I'll pay for it.'

Yes, Barry's rheumatics necessitated monthly steroid jabs, and with these words I'd just sighted up to a life-time's chicken puncturing. Perhaps it was the drink, the *bonhomie* or a sense that moving to the country was still exciting, but the couple decided that yes, their lives would be enhanced by a medically challenged cockerel called Barry. (Or perhaps they reasoned that agreeing to take him on was the best way to make me go away.)

When I told Jane she was pleased at the rehoming, but more or less rolled around on the floor with laughter at the prospect of my administering injections to another male, even if that male was a chicken. I'm not good with needles and for our first decade as a couple Jane and I never went outside Europe because I couldn't bear the idea of inoculations. I was in my mid-thirties with a looming honeymoon when Jane more or less dragged me to the surgery for a single jab.

Now Barry's medical needs meant a trip to our local vet so that I could learn how to jab him, and as the day dawned I was feeling very nervous, so nervous, in fact, that my wife came too. Our German vet Anka had a calm bedside manner which was a little more relaxing than that of the practice nurse who'd punctured me ten years or so before. This woman was short, stocky and in her late fifties, with broad shoulders, a blue serge top, not a lot of neck, and a determined manner. The walls of her consulting room were painted depression brown; the wallpaper was stippled as if the plaster had boils. There were posters of things that had turned red and distended.

'I may look like a grown man,' I said to the practice nurse's back. She was rummaging in a pile of papers for my notes. 'But I am in fact a four-year-old child in disguise. Being frightened of needles is a boy thing, I think.'

'Not just boys,' replied the nurse. 'When my daughter was giving birth she wouldn't have a pain-relieving injection. She doesn't like needles either. She was lying there screaming. I had to give her a little slap and tell her not to be so daft. It was silly really, because she's a nurse too.'

At the time this seemed a little cruel, but with Barry extracted from his carrier and standing placidly on the vet's examination table, Anka suggested something that sounded almost as bad.

'He does not feel much, so we can put water in the syringe, and you can practise on him as much as you like,' she said brightly.

This didn't seem kind, so I asked if perhaps I could practise using an orange before giving the bird the needle, but Anka laughed and said this wouldn't be necessary.

'Well I suppose we'll have to get on with it,' I said in a voice which sounded distant and high pitched. Barry might not have had the faintest idea what was about to happen but I did, and I wasn't looking forward to it. I was rapidly finding that giving a chicken an injection was very nearly as nerve-racking as being given one, or waiting for it, something that 'my' practice nurse had made worse by being unable to find my papers. Time passed, papers were shuffled, my notes refused to be found and I'd thought about making a run for it. Perhaps Jane realized this, and seeing images of a honeymoon in Bognor Regis rather than Cape Town, turned to the nurse and said: 'Look, it's taken me about ten years to get him here. Stick the needle in him now.'

Well, it didn't hurt, much, and afterwards the nurse gave me a sugar lump.

'Polio vaccine,' she explained, before adding a piece of medical advice which I guarantee you will never hear on an NHS helpline. 'As a result of taking that sugar lump you will secrete polio virus through the pores of your skin for the next sixteen days; so don't wipe your bottom with your finger and stir someone's tea with it afterwards.'

I never have, but all that proved was that I'm not the ideal person to give anything an inoculation. Which brings us to the question that you've probably been asking for years: 'How do you give a chicken an injection?' Anka explained that one of the two big wing muscles on either side of the breastbone were ideal for the purpose. 'So,' she said, 'over to you.'

I clasped the syringe, hunkered down, lifted one of Barry's wings and prevaricated. Eventually Jane more or less said 'Stick the needle in him now', and after a little more ducking and weaving, I took the plunge and gave the bird his jab.

Barry seemed entirely indifferent to the needle, although its contents soon put a spring in his step. Back at home Knocker had returned and was looking remarkably chipper as well. Like the other builders, he was working hard in a way which made me feel guilty as I trundled into the garden for an egg-collecting 'screen break' and other distractions. One of these was Knocker's nose. The liquid birthday party appeared to have made it, and the rest of its owner, a little redder than before, but the bridge was deeply scarred and covered with a large scab. I plucked up courage and asked him what had happened.

'Fell over,' said Knocker ruminatively. Then he smiled, something I hadn't seen him do before.

'It was all right, though – I had a soft landing.'

I gave him a quizzical look.

'Gravel.'

Leaving me to wonder if the gravel still bore the imprint of his nose, Knocker went back to work, and I went into the garden. As I'd suspected, Hoover's inadvertent trashing of Slasher's fruit-cage nest area had resulted in her abandoning it. Sadly the bird hadn't returned to laying in the henhouse, and I was once again searching fruitlessly for the place where her eggs were now piling up.

SECOND COMING

Soon afterwards Barry departed to his new home with the couple I'd met at the party. He moved upmarket, to the gardens of a large, sixteenth-century farmhouse, an expanded harem, and a family containing two little girls aged about seven. They were sisters, who to my untrained eye looked identical, and they were big fans of the chickens, which is where the trouble started.

About two weeks after Barry's visit to the vet, their mother rang.

'We've got a bit of a problem with the cockerel,' she said.

'Yes?'

'I'm afraid so.'

I asked what it was and the lady explained that her daughters had a favourite chicken. Unfortunately, it had become Barry's favourite too, and he was keen to enjoy its company. 'The thing is, I'm finding it very difficult to answer the question "Mummy, why is Esmeralda bald?"' Rather lamely I said something vague about how things would probably improve.

When I came to give Barry his next jab they clearly hadn't. Esmeralda's back was featherless, so that it resembled a monk's tonsure, and she appeared to be hiding from Barry in a greenhouse. A children's birthday party was in full swing at the farmhouse, and as I slunk past the kitchen window with a loaded syringe in one hand I could see a posse of little girls, including the sisters, in puff-sleeved party dresses. I decided to be as discreet as possible. A strange man stabbing a pet hen with a syringe might not go down well as children's entertainment.

Keeping a low profile was difficult, because although Barry's rheumatics were giving him gip and he was limping slightly, he was still capable of hobbling at high speed. I was unfamiliar with the garden, which had raised flower beds and a lawn strewn with an obstacle course of children's scooters, balls and toys, so catching him was more difficult than I'd anticipated.

Not wishing to stab myself with a syringe filled with steroids that were never intended for human beings, I left it on a dustbin and did my best to catch Barry as unobtrusively as possible. This involved a lot of fruitless running about, and hearing cheers and girlish giggles every time I got near the house. The building appeared to be modelled on Cold Comfort Farm, with bits that jutted out where you didn't expect them. The edge of the kitchen roof was less than six feet from the ground, and topped off by black guttering.

For reasons of originality, this was not made of cheap plastic but very solid cast iron, which meant it hurt a great deal when I ran into it, glancing my right cheek hard against it, and snapping my head backwards. I staggered onto the lawn, where I bent double and clutched at my face. I did see stars: they were blue on a black background, and resembled freshly blown dandelion seeds. I could also hear childish voices singing 'Happy birthday to you, happy birthday to you'.

I kept repeating the mantra under my breath: 'Mustn't swear, mustn't swear, mustn't swear, mustn't ****ing swear.'

After a while the agony ebbed, a throbbing began, and nobody came out to find out what was happening. Ten minutes later I finally cornered Barry in the greenhouse, did him with the syringe (something which I'm ashamed to say gave me a certain grim pleasure), then drove home for a lie down with some ice cubes wrapped in a towel clamped to my face.

Soon afterwards there was a knock at the back door. Hoover went bananas, his high-pitched yapping drilling into my head, until he rammed his own through the cat flap the better to bark at the builders, who were standing on the other side of it.

I cracked open the door and was greeted by Number One Builder. It was hard to decide which of us looked sadder. Knocker and Male Bimbo Builder looked pretty miserable too. They were staring into one of the foundation trenches.

'We've found a well,' said Number One Builder. 'In the footings.'

I joined the depressed trio, clustered by a trench where our foundations were supposed to go. Hoover had joined them and was staring imploringly up at Male Bimbo Builder, who was clutching a plastic box containing sandwiches and the obligatory Scotch egg. Ever since this job had started, the poor man had yet to eat a whole one, as our dog always made his presence felt at dinnertime.

Hoover wagged his tail appealingly as we looked at the well. This was almost a cliché. A perfect, circular brick hole, the sort a small child might draw.

'If it was ten foot further in, we could probably have preserved that,' said Number One Builder, 'but it isn't and it's got to come out.'

I asked a predictable question. 'How long will that take?'

Number One Builder pulled a face, and exhaled in the depressed way Hoover has when he feels he should be the centre of attention but isn't. From this I gathered that the timeframe would be an elastic one.

'The trouble is we'll have to get a structural engineer in to say what we have to do, and there's not much we can be getting on with until that's sorted out.'

The inevitable question about how much all this was likely to cost produced a slightly hunted look. The ensuing silence was broken by a rhythmic thumping noise. Hoover had sat down ostentatiously and was banging his tail on the ground in a *'Look at me I'm being good'* kind of way, as his eyes bored into the side of Male Bimbo Builder's head. The dog was sending out *'Give me your lunch'* thought rays. Hoover has an instinct for finding a human soft target when it comes to food, which is perhaps why he hadn't tried the same tactic with Knocker.

'Anyone for a cup of tea?' I asked. It seemed a good way to break the impasse.

Back in the house the phone was ringing. I picked it up, feeling my sore face throbbing against the handset as a woman's voice said, 'Hello, I'm the lady who's got Barry. I wanted to see you this morning, but what with the party I couldn't get away. I'm phoning to say that I'm sorry, but he's got to go. When can you collect him?'

*****

In the builder-less hiatus that followed, the livid bruise on the side of my face started to subside. Barry was once more in the fenced-off area behind our summerhouse, and in rude, vocal health. He seemed keen to resume remote hostilities with Svenson, who was once again confined to barracks and not pleased as a result.

'We've got to do something,' said my despairing wife during a brief let-up in the vocal duelling.

Then we had a stroke of luck. Chris, a near neighbour, mentioned that he had a ten-strong flock of Marans chickens.

'I don't suppose you want a Marans cockerel to go with them, do you?' I asked, more in hope than expectation. To my amazement he liked the idea, and was not put off by Barry's chronic medical issues. Not wishing to waste time I offered to bring the bird round as soon as he was ready.

'No time like the present,' said Chris.

Feeling just a little smug at managing to rehome a cockerel twice, I deposited Barry with his new harem (whom he seemed very pleased to see) and said I'd be back in a couple of weeks to give him his medication (Barry, not Chris). Sven and his girls were obviously pleased at having the run of the garden again, and things quietened down a lot – although Chris lived sufficiently close to us for Sven to just about hear Barry and vice versa, so there were still the occasional vocal sparrings.

'Still want some, then?'

'Pardon?'

'I SAID, STILL WANT SOME THEN?!'

A week after Barry's second departure the builders were back, accompanied by a man in a suit, clutching a clipboard and sucking a pencil. This was the structural engineer. I made everyone tea and hovered in the background, feeling somehow that I was intruding, despite this being my home. After a series of low-voiced exchanges and pointings the man departed with a promise 'to let you know what I think's needed here'.

After he'd gone Number One Builder asked, 'Do you want someone to shoot those bloody rabbits? You've got a lot of the little bastards in your garden. Whenever we look round we can see them running about.'

By now it was lunchtime, and he and Male Bimbo Builder had retired to a garden bench. The sun was shining and a mild warmish breeze was blowing across the garden, bringing with it the indistinct prattle of Radio 2 from a paint-spattered ghetto blaster. Its owner reached for a Scotch egg, stopped and looked into the eyes of a dog who had turned pathos into an art form years before. Hoover sat at the man's feet, regarding him with a steady, sad gaze.

'Hoover,' I said. 'Bugger off.'

The dog retreated about ten feet, stopped, sat and looked even more professionally miserable.

'Is it all right if he has a little bit?'

I sighed. 'If you want to, but he's not supposed to beg.'

Seconds later our dog was at the man's feet, head up, jaws wide open, and a quarter of Scotch egg was dropped into his bearded maw. It barely touched the sides as it went down.

Number One Builder laughed. 'You've started something now,' he said as Hoover reconfigured his features to once again look sad. Ignoring this I asked Number One Builder about bunny murder. The rabbit problem we'd had in the winter had progressed from an occasional irritant to a minor plague, as they'd taken to breeding. Although the chickens were no longer sharing territory with Barry, the bunnies were taking up the

slack. I'd started seeing small obscenely cute baby rabbits gambolling across the lawn, something the hens
seemed to find entirely uninteresting. Number One
Builder had seen them too, and viewed what he saw
unromantically as pests that had the potential to make
themselves useful as a source of food. We also both
agreed that they would be a magnet for foxes.

'Do you mind if I bring my air gun then? Would
tomorrow be OK?'

Steeling myself for *lapin* infanticide I nodded, knowing that a long-awaited early-morning physiotherapy
appointment would get me out of the house, and I
wouldn't be around when it happened.

BACK AGAIN

It was 7.30 a.m. the next day and the rabbit cull had yet
to begin when the phone rang.

'I've got them doves for you,' said a voice.

It took me a minute to realize that this was the farmer
who'd promised Jane some wild doves. His timing
wasn't great, as I'd been clutching my car keys and
heading for the door and my bout of NHS physiotherapy. I explained the situation.

'Could I come at the weekend?'

'Not really,' said the farmer. 'I've got them in a bag.'

'I'd better come now then.'

'I think you'd better.'

I put our collapsible dog cage (bought for transport-
ing chickens – obviously) in the back of the car and

headed for the farm, a rather wonderful organic place called Cuckoo's Pit (but inevitably known to us as Cuckoo Spit). When I arrived the farmer greeted me warmly and handed over a writhing brown sack.

'Would you like some more?' he asked.

'Why not.'

'You wait there,' he said, disappearing into a barn and shutting the door behind him. What followed was the sound of a man in heavy work boots running around and cursing in a dark space, a lot of flapping and avian 'woo-woo-wooing', followed by the occasional crash. After a few minutes the door was cracked open and a pair of hands clutching an unhappy-looking dove emerged.

'There's one,' said the owner of the hands.

I somehow managed to clasp the bird and hold the mouth of the sack before forcing the dove into it, rather, I imagine, as one might squirt minced meat into a sausage skin. We repeated the process three or four times, when the farmer popped his head round the door and said, 'You should have eight birds in that bag. Want any more?'

I said eight was just fine, and we gingerly tipped the bag's thoroughly ruffled occupants into the dog cage. We could see then that what we had were six doves, one half dove/half pigeon, and what appeared to be a homing pigeon, with a little orange tag on one leg. My inclination was to let these last two birds go, but thinking that this might not go down well with the farmer – who'd pronounced himself delighted to be shot of them – decided to review the situation at home.

A series of spasming twinges at the base of my spine reminded me that I had a fast-approaching appointment with an NHS back tweaker, and couldn't hang about. Throwing an old blanket over the cage I headed for home and decanted it into our garage, where my wife proffered our new arrivals some grain. They began eating without ceremony, although they did jump when there was a distinct 'Crack!' sound from the garden. Number One Builder was wasting no time in dealing with the rabbits.

*****

Our local hospital is a huge, grey edifice dating from the seventies. Just being there, walking past lines of people standing outside smoking furiously (despite in some cases being plugged into drips), made me feel a bit peculiar. I queued up for some inconclusive manipulation and left with a sheet of paper filled with slightly sinister-looking stick men depicted at various angles (that would, I suspect, be illegal in Texas – at least with the lights on).

'Exercises,' said the physio. At least I had some instructions on how to look after my vertebrae, but as I drove home I realized that I could really do with some when it came to dove husbandry. The information sheet given to us by the bearded, khaki-wearing dove breeder had vanished, and our first attempt at keeping the birds hadn't ended well. I really didn't know what I was doing.

When I got home the builders had gone, and Jane was feeding the hens in an uncharacteristically rabbit-free garden. Number One Builder was apparently a capable shot, and had described to her a dispatching which made our rural home sound like a murder alley. Still, if this discouraged further bunny visitors, at least for the time being, its attractiveness to foxes would also be reduced. My wife was on her haunches, and told me this whilst holding a handful of feed towards Brahms, the stolid, shy Brahma hen, who came at it in a careful arc, pecked diffidently, then scuttled away. As we watched her my wife said, 'It's time we did something about putting the doves to bed.'

We had been unusually organized in preparing for a second flock of doves, and had got our hands on superior accommodation for their confinement. No more flapping gauze for us and the doves to get tangled up in. Instead we had three wooden boxes, sort of bird prisons, which could be attached to the outside of the dovecote. Their captives would have more room, and food bowls wouldn't get knocked to the ground every time I tried filling them up. So that evening we attached the three box arrangements to the dovecote, put food and water bowls in each. Then we tottered down the garden with the dog cage and its cargo, which collectively was not best pleased at being moved, then pondered what to do next.

'Which ones are boys and which ones are girls?' asked Jane. Neither of us had the faintest idea.

*****

For the chickens the new arrivals were potential sources of strife – or possibly takers of food that by rights should go to them – so a posse of grumpy hens soon surrounded the cage and eyeballed the occupants, which did not improve the atmosphere.

'Go away!' said Jane. 'Go on, shoo.'

Chickens are good at going away and then instantly coming back again, which is exactly what ours did. This was followed by more close, unfriendly scrutiny of the doves, who stared up at the giant, predatory chickens with understandable alarm.

'I think we should just get on,' said Jane.

We decided to pair the doves using size as a determiner of sex, which meant two birds per prison, and putting one big and one smaller bird in each cell. We chose to hang on to the half dove and the racing pigeon for the time being, and had rigged up a smaller piece of mesh over the pop hole through which they'd be posted.

Jane handed me a bird, and I wobbled up the ladder, clutching it close to my chest, feeling its tiny heart beating like a steam hammer and hoping it wouldn't pass on any mites or ticks as a parting gift. When everyone was installed in his or her new accommodation there was quite a lot of cooing. Meanwhile the chickens were trying to break into the now dove-less dog cage in the hope of finding yet more things to eat.

'That's sorted that out,' I said, displaying a touching *naïveté*.

The complete inaccuracy of this remark quickly became apparent. Sounds of violent scuffling and angry cooing soon started coming from the dovecote. We'd look up at one of the prisons and see a wing appear in a doorway, and then vanish, as its owner was dragged out of sight by its room-mate. Size, apparently, wasn't everything, and some of the smaller birds seemed to be getting the upper hand with these skirmishes. What it meant was that we hadn't got the correct male/female ratio, which forced me back up the ladder to keep swapping everyone round until the punch-ups ceased. (To describe doves as 'birds of peace' is clearly another sad human fantasy; only the half dove and the racing pigeon seemed content with each other's company.)

All this mucking about took around half an hour, but I could see that leaving things as they were wasn't an option. Birds are very territorial, and forcing two males into close proximity was cruel and could lead to injury or even worse. It turned out that we'd been lucky, in that there was a 50/50 male/female ratio with the white doves. We just lacked the skills needed to work out which was which. It was a case of swapping everyone around and waiting for the yankings, batterings and gratuitous violence to stop.

'I feel almost as exhausted as they do,' I said, as we tramped up the garden and back to the house.

'You need a drink,' said my wife.

'Do I?'

'Oh yes.'

*****

Two weeks after the new doves' arrival everything still appeared sweet. For allegedly wild birds they seemed remarkably tame, and peered at me with interest rather than the abject terror of their predecessors when I climbed up the ladder to feed them. The sun was shining and my back felt great. Although I had some pressing journalistic work, I first had to wander round to Barry's new home and give him his jab, so having loaded a syringe with steroids I made my way there to be met by Chris's wife, who, slightly confusingly, is also called Chris.

'Barry is very pretty,' she said. Her tone was friendly, but neutral. 'The thing is, my husband is supposed to be getting rid of stuff, not acquiring it, and [quite a lot of significance was put on the word 'and'] when he does, he's supposed to ask me first.'

Apparently Chris had not discussed Barry with Chris.

'So we've decided that Barry can stay here until you find him a new home,' she said.

'You've no idea the trouble you've caused,' I said to the wriggling patient, as I administered his medicine.

Initially I felt rather stumped. Second-hand cockerels are not popular, thanks to a mix of noise, aggression, and a lack of eggs, so finding a home for a faulty one three times running was not going to be easy. What the hell was I going to do about Barry? Was this how the Ancient Mariner felt about the Albatross? For several days I was stumped, and the return of the builders with a missive from the structural engineer proved to be a further distraction.

'Sorry,' said Boss Builder, with the look of a small boy who has just accidentally crashed into a priceless china vase and smashed it to bits. When someone starts a sentence with the word 'Sorry' you know the rest of it isn't going to be cheering. What came next wasn't.

'He says that we have to take out the well, dig down another metre, and put in steel reinforcements. It's going to take us a couple of weeks.'

I looked at the well and cursed it, then asked, 'How much will it cost?'

Boss Builder's shoulders twitched upwards so that he looked as if someone had stolen his neck. Rolling his eyes like a spanked puppy he said, 'About another £3,000.'

Later, as I contemplated this cheerful news – and wondered how Jane would receive it – I had an idea about rehoming Barry. It would add to our financial burden, but if it worked, the expense would be worthwhile. We live in a part of Kent graced with a weekly advertising free sheet called the *Wealden Advertiser*. I phoned and spent £18.50 on an advertisement with the tantalizing headline 'Slightly defective cockerel seeks home'. A week later, when I'd all but given up hope, the phone rang.

'Hello,' said a well-spoken male voice. 'I'm phoning about the cockerel. I run a shooting range near Biggin Hill.'

Realizing that as an opening gambit the bit about the shooting range might cause some apprehension, the owner of the voice added, 'but it's all right, we don't want to use him for target practice.' It turned out that this man kept yet another flock of Marans hens, which lived behind a large black farm building, and he reckoned a boyfriend might put some spice into their lives.

'The place is run by a retired gamekeeper. He's a good bloke. Show him how to give the bird an injection, and he'll do the rest.'

This seemed like the ideal solution. Barry would have yet another set of girlfriends through which he could pass his rheumatic genes. He'd be living a long way from us, and somebody else would be giving him his jab.

A handover was arranged, and a few days later I drove to a shooting range fairly close to Biggin Hill aerodrome, but within strafing distance of the junction of the M20 and M25 motorways. The gamekeeper emerged from a battle-scarred Toyota pick-up. Aged about seventy-five he had red veins in his earlobes, a bulbous nose, and the sort of complexion that implied its owner had been outdoors every day of his life since the winter of 1963. He wore a flat hat and a sour expression. This man clearly wasn't pleased to see Barry. Nor was he particularly enamoured with me.

I followed his Toyota to a large barn, behind which was a run filled with hens busy scratching about. Barry seemed much more interested in them than anything else as I showed the gamekeeper how to administer the bird's jab.

'Hmmm,' he grunted unhappily. 'When I find a sick bird, I don't normally give it an injection.'

Introduced to a third set of girls in as many months, Barry looked far from sick. Down went his left wing, making a fan shape, as he did a little circular dance to indicate that Mr Love Pants had arrived. That, I'm glad to say, was the last I saw of him, but it was nice to part on a high.

To this day I regularly drive along the M20, and often think about Barry. You may also find yourself on the same stretch of road, not long before it joins the M25, heading towards lovely Clackett Lane services. There is quite often a traffic jam, which will allow you to view the scintillating scenery. Should you be stationary and turn to your left, you will see, about a mile away, a big black barn, behind which, with a bit of luck, is a lightly limping cockerel having a good time.

Back in our garden, summer was now rolling over the horizon. The days were getting a lot longer and the light had changed in a way that is hard to describe, but is presumably connected to the planet getting closer to the sun. Svenson was getting very randy, the doves were getting restless and keen to get out into the world. It was almost time to open their prisons and find out if they would stay with us or decamp *en masse* back to Cuckoo Spit Farm. Before that we would have to do something about the lost racing pigeon and its dove/pigeon companion, as we weren't keen on them hanging around to bonk or be bonked by our pure-bred doves and muck up the strain. Call us avian eugenicists if you want, but no apologies. We wanted dove-coloured doves.

We didn't want rabbits, and although we hadn't seen any since Number One Builder had blasted the last lot with his air rifle, the lawn occasionally showed evidence of their berry-like droppings. This we knew was not the end of the story, and more murder lay ahead.

So did the imminent arrival of the Bricklayer who would start building the extension. Having been given the OK to rip out the well and deepen the foundations, the builders got on with the job, working from first light to dusk in a bid to make up the lost time, and performing something akin to construction root-canal work, buttressing up what was now a very deep trench with steel lattice. The garden echoed to the sound of blokes chatting and guffawing, crowing, clucking, cooing, Jeremy Vine on Radio 2 and the satisfying rumble and slurp of a concrete mixer. Soon after that the foundations were filled with concrete, and for the first time you could discern the shape of the building that would eventually stand on them. Jane was ecstatic, and even I felt a bit excited at the prospect.

Despite the presence of the builders, who were charming and whose work ethic put mine to shame, everything else in the garden and on the domestic front seemed pretty settled. A distraction-free summer beckoned.

SUMMER

DUCKING THE QUESTION

'Ducks?'

Jane gave me one of her winning smiles.

'Well, we have been talking about it for a while.'

This was true, in that she'd bring up the subject of duck ownership, we'd have a discussion about it, I'd demure and nothing would happen. Now we were talking about it again, and my lack of enthusiasm must have been obvious.

'We've only just got rid of Barry, the doves are still in prison, we've got a house filled with builders and – with the possible exception of Knocker – Svenson is trying to bonk anything that moves. Is now really the right time for more animals?'

My wife gave me a coquettish look and a copy of a small ads newspaper, opened at the 'Pets and Livestock' page. She'd ringed several advertisements.

'Indian Runner ducks. Two pairs, £35 each' ran one. I didn't know much about ducks, but had a feeling Indian Runners were the long-bodied, wine-bottle-on-legs variety, something Jane confirmed.

'They don't fly. Well, they don't fly much, so you wouldn't have to clip their wings to keep them in the garden,' she said brightly. My beloved had clearly been doing her research on the quiet, the better to soften me up with the life-enhancing aspects of duck ownership.

'Why wouldn't you clip their wings?' I asked.

'I don't need to and anyway, I don't know how,' was the rejoinder. 'And it's your job.' This was said as if it was blindingly obvious.

I decided to ignore that bit and asked, 'How will they get on with the chickens?'

'Fine,' said my wife. 'The lady who's selling them says hers grew up with chickens.'

'So you've already spoken to her?'

'Well,' said Jane, who was doing a good job of not making eye contact. 'I have a bit. It wouldn't hurt just to have a look, would it? We've nothing else on this weekend.'

If we did end up sharing our lives with some ducks, I knew they would have to cope with something approaching a chicken-hormonal war zone, driven by Svenson's increasingly unattractive behaviour, which seemed to be getting worse as the weather improved.

Although chickens are the acme of domestic, or domesticated, animals, they are far from immune to wild urges and survival mechanisms. For Svenson the long early summer days gave him plenty of time for increasingly frequent sexual encounters that had all the intimate delicacy of rugby tackles. Bella, Nude and

122

Meringue were his favourite conquests, something which gave none of them any obvious pleasure. Showing that he was also happy to keep things in the family, the bird had taken to jumping on his mother Ann and sister Ulrika. Slasher and Too were harder to catch, and so suffered less molestation. Squawks was easier to keep up with, but somehow more resistible.

Poor old Brahms was apparently completely beyond the pale, so she wasn't getting bonked, but increasingly she wasn't getting fed either. Mealtimes had become a battleground, because Svenson kept chasing her from the grub he wanted his desirable girls to get. This isn't uncommon in chicken society, but none of the other male birds we've owned has come anywhere near the blatant Svenson in this behaviour. So Bella, Nude, Ulrika and of course Meringue were being positively encouraged by him to eat, whilst the lesser birds were often harried and bullied.

Chickens do genuinely have a pecking order, and being pecked is exactly what that entails. Think of the classic *That Was The Week That Was* sketch on class featuring Ronnie Corbett, Ronnie Barker and John Cleese ('I'm working class, so I look up to him', 'I'm middle class, so I look down on him,' and so on), and you pretty much get the average chicken flock's dynamic (but with added violence).

Unless a cockerel is extremely wimpy or young, he will be in charge, with everyone else slotting in to their allotted places. These dictate who gets to eat first, and often most, and the birds who weren't Svenson's favourites frequently received a jabbing peck if they tried piling in before he thought the others had had their fill. Slasher was often on the receiving end of a swift *'Get off'* peck, but for Brahms it was usually worse, especially so in the summer, the time when chicks are hatched. She was never going to be a parent and, instinctively, Svenson regarded her as superfluous.

So at feeding times I was having to stand guard to make sure she wasn't excluded, but I'd been slow to notice how often she'd been virtually going without. With a lively flock of birds there's usually a lot of distracting stuff going on, and things are easy to miss. Brahms would make a quiet approach to the food, often lowering her head when she arrived and holding back, then taking whatever she could until Svenson or a senior bird chased her away. This was a tactic also pursued by Squawks, but often with more success. When Svenson spotted Brahms he would always charge, forcing her to run quite some distance from the food. I was a little shocked when I picked her up, and found a skinny bird, with easy-to-feel, fragile bones just under thin, wrinkled skin.

Sickly chickens don't like to make a fuss, because this draws attention to their frailties, and the rest of the flock won't look out for the weak: quite the reverse. They're a liability to everyone else, so Brahms had once again been slowly fading. Being an introvert, she was very reluctant to eat from my hand when food was proffered. She would just fix me with an unmoving, sad gaze, then lumber away when other, greedier birds turned up and helped themselves.

Some hens are quite relaxed about being handled, and come to associate human contact with something to eat. Others grow to actually like the attention. Brahms wasn't one of these, but if she wouldn't be hand fed, particularly when being held, a solution was needed, as I could see that she risked being starved if this continued.

Going for the cockerel when he went for her made me feel better, but wasn't a cure either. When doing this I found the most effective method was to pretend to be a very large chicken, flapping my hands up and down as if they were wing feathers, and running after Svenson shouting things like 'Go away, you git!'

Since we have high, wooden fences none of the neighbours could actually see this, which was probably just as well, but presumably they can hear a lone man, apparently talking to himself. Then there were the builders, and in particular the Bricklayer, who was building the shell of our extension at a furious pace. By the middle of May he and the others had the structure to waist height, scaffolding had been erected, and they'd continued working upwards. Of course that scaffolding meant they had a panoramic view of the garden, something I'd forgotten more than once when flapping around after Svenson. Nothing was said, but the significant silence that sometimes greeted my return spoke volumes. They probably thought I was nuts.

My efforts were only partially successful, as they would cause Svenson to semi-bugger off, then come back and start the bullying all over again. Chickens have a talent for monomania, especially where food is concerned, and Svenson's DNA was telling him that Brahms had to go. Jane and I weren't having that. We were especially fond of Brahms, whose quiet presence had become a distinctive feature of the garden when we were working there. Some of the birds would make a show of turning up, bustling about and actively getting in the way, but Brahms just appeared, keeping an eye on us, and we felt protective towards her.

The plan we hatched to help her at mealtimes was simple, but it worked well. We tipped most of the food into a feeder in one chicken run, but had a bowl ready in the second one. As the birds made for the primary food source, we'd gently chivvy Brahms to the other and shut the gate. To start with she found this alarming, assuming that we were trying to catch her, but after a couple of weeks she'd got the message and began actively colluding with us, keeping her distance, arriving discreetly and then running into 'her' aviary. Svenson would still try heading her off, but the threat of a boot up the backside – never actually carried out – was enough to distract him. As Brahms began feeding regularly and became gratifyingly plump, we noticed that Meringue, and occasionally Squawks, had picked up on the subterfuge, and would fall in behind her as uninvited, but not unwelcome, guests.

*****

Would there soon be more guests at mealtimes? After the chicken-feeding ritual, and once I'd given the doves their breakfast, I'd returned to the house to find that my wife was still mentioning the 'd' word. Ducks were on her mind and she would not be fobbed off. So after some further prevarication on my part, persuasion from Jane and a telephone call to the people selling Indian Runners, we had strapped Hoover into the back of the car and were en route to where they lived 'just to have a look'.

Jane and I have a long-established understanding when it comes to navigating, which she generally does, because she's better at map reading than I am, and is quite happy to be driven. So the three of us trundled along sunken country lanes that were lush, green culverts, often little broader than our car, hoping we wouldn't meet anything coming the other way. Then we'd reach a junction or crossroads. Jane would squint at the map and make a sort of flapping hand gesture and say something useful like, 'This one! This one! That one,' as we sailed past a turning. 'Sorry,' she'd add, as I'd stop, huff, back up to the junction and point the car down the correct lane. Jane does know the difference between left and right, but her brain won't always let her say which is which. When she was a primary school teacher one of her abiding worries was having to take games lessons, especially football matches, for fear that she'd create mass collisions as she tried to make children go in the right direction.

So I have come to recognize her hand gestures. If the one nearest me is thrashing, the car needs to turn right. Something jiggling at the very corner of my left eye's peripheral vision, we need to go left. So, after a mix of frantic gestures and cries of 'This way!' we finally turned into the yard of a large, immaculate stud farm, to be greeted by a tall, very elegant woman in her forties. She had a look of casual, easily worn wealth, and was dressed in the sort of clothes you'd associate with someone who spent a lot of time with horses; riding boots, jodhpurs, a tasteful, expensive-looking top.

All these things were remarkably clean. Not, I suspect, because she avoided mucking out, but had the capacity to avoid getting covered in guano. This is a skill I've never learned; in fact, it's been the cause of some discord in our house. I clean out the birds at least once a week, but don't always remember to change into the sort of clothes that don't matter if some of the filth gets onto them. As a result I've returned from guano shovelling to be greeted by cold looks and comments along the lines of 'Those were your good shoes.'

In the stud farm's yard a small dog jumped around its owner's immaculate boots, causing Hoover to wipe his nose on the car window as he squashed his face against it to get a better look.

'Have you come about the ducks?' asked the equine lady.

We said yes, and were led to a concrete yard surrounded by stables.

'They're in there,' said our guide, pointing to one of those split stable doors that allow four-legged residents to poke out their heads. We did the reverse, breathing in the agreeable scent of hay and horse. Looking down we saw four upright shapes scuttling about. Two pairs of Indian Runners, the males with markings not unlike Mallards, with green heads, and white plumage enlivened with blue and brown feathers. The females were a pale, speckled brown. All of them were pretty but ungainly. One pair seemed to find us particularly interesting, probably because they thought we'd feed them. I wouldn't say that we bonded, but we did reciprocate their interest.

'We like those two,' said Jane.

'Yes,' said the woman. 'They're very friendly. My daughter likes them too, so she'll be sad when they go, but she knows they're for sale.'

Would we be the ones to take them away?

'We are going to, aren't we?' I said to Jane.

'Looks like it.'

We told the vendor that we were interested, but ill-prepared, and would have to fix up temporary accommodation, on the basis that if she didn't like the idea of the ducks going to a pair of novice owners, she could hang on to them.

'They like water to swim in, but they don't actually need it, so if they're secure and have space to run about they'll be fine,' she said.

This seemed rather unkind, and some sort of duck paddling pool was added to the mental shopping list of stuff we were going to have to buy. Then Jane fished out her purse and I went back to the car to collect the carrier. When I got back the daughter made an appearance. Aged about fourteen, she looked like a smaller version of her mother.

'Are you buying the ducks?' she asked. We nodded.

'Oh dear,' she said, sounding genuinely upset, but fortunately the tearful parting I'd half-expected never materialized. The girl gave a teenaged 'Well, that's it then' shrug and left us to it. In the space of a few minutes she'd done 'being upset' and had moved on.

We let the owner catch the ducks. She slipped into the stable and rapidly picked up the drake, who cycled his stumpy legs, trying to yank manicured fingers from his midriff with the claws of his orange, webbed feet. She deposited him in the cage we'd brought, and he began quacking anxiously, but the woman was soon back in the stable, scooping up his mate, and when the pair were reunited relative calm ensued. After that we covered the cage with a travel blanket, and its occupants fell silent.

The drive home was punctuated by a visit to a garden centre where we bought some metal garden stakes, rolls of plastic fencing, and a pair of heavy dog-food bowls that the birds would find difficult to knock over. Then we detoured via a pet supermarket and bought some duck food.

'This really isn't how we should have done this,' said Jane.

Once home we placed the cage in the shade, filled the bowls with food and water, and offered them to its nervous occupants, who overcame their fear to eat copious quantities of something called 'Floating Discs'. Oddly, the hens seemed entirely indifferent to these comings and goings, choosing instead to lounge about, take leisurely dust baths or peck lazily at the lawn.

I'd had an idea for temporary duck accommodation. We'd bought some plastic composting bins that looked like giant grey pepper pots, or embryo Daleks. If I cut a tall doorway into one of them, and mounted it on a solid wooden base, it would make a very serviceable residence for a pair of Indian Runner ducks. I spent the afternoon with bits of plywood, hinges and shoot bolts. Often my adventures into DIY land end in bloodshed and blind fury, and the creation of something wobbly and semi-useless. For some reason this particular excursion didn't take me up a blind alley, and by seven in the evening (although I was ready for a drink) I was the author of a solid, if slightly eccentric, duckhouse. Three large shoot bolts secured the composting bin to a wooden base that a fox would find very difficult to upend. I'd kept the sliding panel at the front, but cut a doorway above this and fitted a hinge, so the tall Indian Runners could get in and out. To make this easier still there was a wooden ramp, which worked like a draw-bridge and gave further protection from the outside world.

Necessity had been the mother of invention, but the end result looked far more pulled together and well thought out than its *ad hoc* creation would lead you to expect. Jane had fenced off the part of the garden where the ducks would live, and we'd found an old plastic container that had once held water plants. It looked a bit like a replica of a Victorian tin bath, and would serve as a swimming pool. With a length of timber functioning as a ramp, we filled it with water and, as night fell, posted the wriggling ducks into their temporary home and locked them up for the night. After some brief desultory quacking they fell silent.

DAY RELEASE

In the dovecote things had become more settled too, and we were contemplating letting the occupants out. They seemed thoroughly institutionalized. They all had healthy appetites, and were good at knocking over their food containers, often scattering the contents on the ground, where our chickens were more than happy to ensure that it didn't go to waste.

We'd already decided to release the semi-dove and try to find the racing pigeon's owner, whom we reckoned might well be missing it. A few days after the ducks arrived, I extracted the semi-dove, said goodbye to it and put it on our wild-bird-feeding table. It was so surprised that it sat there for about thirty seconds before realizing that this wasn't compulsory, and that it could do what it liked. This was to fly onto the eves of the neighbouring house where our first lot of doves had previously taken shelter. It perched there for a few minutes, eyed its surrounding with disfavour, then flew away in the direction of Cuckoo Spit Farm. It never came back.

The racing pigeon was more of a problem. We felt reluctant to simply let it go, as it apparently belonged to someone, and we wondered if that someone would be pleased to see it again. I took down the number on its leg tag, got on the Internet, and in due course found the bird had originated from somewhere near Southampton. I'd gleaned this from a pigeon fanciers' society, and was given a phone number. My call was answered by a lugubrious male voice which said, with a distinct lack of enthusiasm, that the bird sounded as if it was one of his. We'd talked round this for a bit when he suddenly changed his mind. No, it wasn't his pigeon, but he had a pretty good idea who did own it. Papers were shuffled and another name – which sounded a bit Central European – and a new number were offered. Hoping for a more enthusiastic response I rang it. After a while somebody picked up the receiver.

'Yes?' said a heavily accented male voice.

'Have you lost a homing pigeon?' I asked.

'Yes,' said the voice.

'I think I may have it.'

'Yes.' This was said in the manner of someone who was used to complete strangers ringing up to proffer lost pigeons. I read out the number taken from its leg ring.

'Yes,' said the voice. 'That's mine all right.'

Its owner seemed no more pleased with the news than the man who'd given me his number. This was confirmed when he added, 'Would you like to keep it?'

This wasn't exactly the response I'd been hoping for. I have a mild St Francis complex when it comes to pets. Why didn't this man want his bird back?

'Well, he's not very good at getting home, is he?'

This seemed a fair point, but I didn't want it either. It was a proven dove fraternizer, and if it got into bonking our flock the result would be more dove-pigeon love children. So I said thanks, but no thanks. After a pause I was asked if we'd mind holding on to it until the owner could come to collect it, which didn't seem unreasonable.

'When could you do that?' I asked.

'In about four months,' came the reply.

This was not what I had in mind, and I said so.

The pigeon fancier sighed deeply. 'Well I dunno,' he said. 'I could always ask Frankie.'

He took my phone number, and promised to ring back once he'd spoken to Frankie from Gillingham. I was only slightly surprised when he did.

'Frankie says yes, but you gotta go to him.'

I asked if it was possible to meet Frankie halfway.

'No, I don't think so,' came the reply. 'He don't like going outside. The most he drives is to Tescos.'

Fortunately – I can't think of a better word, but use it advisedly – I was doing some work for a magazine based in Swanley. If you live in Swanley, apologies in advance, but it is not conventionally beautiful, and the prospect of spending extended lengths of time there gave me feelings of modified rapture. Being paid to do so was entirely congenial, however. Also, making a detour en route to the glory that was Gillingham wouldn't be such a fag really, as one place virtually backed onto the other. So, with the pigeon inserted into a smallish cardboard box with air holes, I drove along the busy, aggressive M2 motorway, and into a depressing, grey hinterland of slightly tatty outer London suburbia.

Frankie lived in an Edwardian terraced house perched by a busy arterial road. The original sash windows had been replaced by featureless plastic ones and the building rendered in corpse-grey pebbledash. The tiny front garden was decorated with a couple of bins and a lot of dead leaves. I rang the bell and heard a brief scrabbling of pigeon claws on cardboard, then the door opened a crack. Behind it I could make out a small, middle-aged man with a creased, careworn face.

'I've brought the pigeon,' I said with a brightness I did not feel.

Frankie said 'Thank you' with an equal lack of enthusiasm and a pair of large hands emerged through the gap between door and frame. They took the cardboard box, withdrew as muttered pleasantries were exchanged, and then paused. Their owner then mentioned that there was now some doubt over who actually owned the bird, as a second pigeon fancier had decided it might be his. Then the door was quickly closed.

Walking back to the car I felt bad. After all, I'd been the cause of this bird leaving a nice life in the country, a plentiful supply of food and an endless succession of blonde girlfriends, for incarceration with a Gillingham agoraphobic, before eventually being reunited with a monosyllabic bloke in Southampton who didn't really want it back (assuming that it was his to begin with).

'I should just have let it go, but it's too late now,' I thought as I climbed into the car and headed for Swanley.

Back home the remaining doves were apparently enjoying loft living. Eggs were being laid and there was a fair amount of bloke-ish strutting, puffing up of chest feathers and a lot of the head nodding and going round in circles dances that pigeons and doves engage in when they want some love action. These were usually accompanied by a great deal of amorous cooing.

The birds also began making a racket when they heard me in the garden. They would emerge from the dovecote into their little wooden prisons and give me hard, pointed looks from behind the mesh walls. Like the chickens, they had come to associate me with being fed, and food was something they could never get enough of.

So I would trundle up and down the ladder to proffer grub. The doves would still beat a retreat into the dovecote itself, but I had the feeling that this had more to do with instinct than being actively frightened of me.

One of the females, who'd laid a number of eggs, was now sitting on them firmly, head down, bum in the air, staring fixedly at me as I emptied and refilled food and water bowls. All this boded well for their release. The first lot of doves had barely laid a single egg, and had little territorially to make them stay in our garden. They'd also been far more skittish than their supposedly feral replacements.

After one morning's feeding session I became aware of frequent, high-pitched squeaking noises from the dovecote, and realized that we now had a flock of seven doves, since one of the eggs had hatched. I clambered up the ladder and peered into one of the prisons and saw, briefly, a small pink thing with floppy folds of skin and a long, scrawny neck, on which was a smudge of head. A knobbly orifice that was clearly going to turn into a beak was open, and the whole thing was slightly Davros-like. This was the first time I'd seen a baby dove.

The tiny creature was straining upwards and vibrating. It wanted Mummy and Daddy to regurgitate nutrients down its throat, but on seeing an unwanted human visitor Mummy clamped herself over her beautiful *baybe*, puffed up her feathers and gave me the evil eye.

*****

Outside, everything was in bloom. Buds were releasing flowers, the grass had begun its irritating habit of growing, forcing me to mow it, and the wild birds were more or less lining up outside demanding that we feed them. Before the builders arrived we'd grown used to dawn raids of squabbling, argumentative starlings, and two nesting pairs of robins, who had territorial spats and would otherwise dance around on the branch where we hang the feeders (over which they also fought). These little birds were often almost inches from our noses. Nature had pressed the fast-forward button, and everyone was assiduously taking everything they could get.

Svenson continued to try proving that he was now a fully paid-up member of the Love God Club, which resulted in an endless round of avian ravishings. This meant two things. One was that I would have to clip his spurs regularly, and the other involved digging out the chicken saddle. This is a hefty piece of shaped, yellow canvas, designed to protect a hen's back from a vicious spur jabbing. It's attached by a pair of elasticated wing straps, designed to accommodate the fuller-figured chicken.

The bird who normally ends up wearing it during this warmer, more passionate time of year is Meringue, since Svenson has a near Wayne Rooney appreciation of older females, and they're always the ones who pay for his attentions. His spurs jut from the back of his legs like a pair of scythes. Tapered to sharp points they can do a lot of damage. I've seen birds with painful-looking gashes, and we've had occasion to keep Svenson in one of the chicken runs whilst his girls went walkabout. This state of affairs caused him to go bananas with a fizzing mixture of lust and boredom, but at least gave his ladies some respite from constant harassment. Although every inch of cockerel was quivering with filthy intent, I always felt sorry for the bird, as he watched his non-conquests ambling about in the sunshine.

The saddle at least meant that his favourites suffered nothing worse than regular indignity and mild discomfort, rather than injury. However, persuading Meringue into it was trying for both of us. She seemed to recognize it and beat a hasty retreat when she saw me coming. Now when I headed for the garden to attach this piece of unwanted avian couture, I hid it in my pocket first.

Meringue might not have the lightning responses of Slasher but she's no slouch, and if I don't have a suitable treat such as tinned sweetcorn – which any self-respecting chicken would impale herself on a rotisserie to enjoy – she will always clear off in such a way that I end up lumbering round the garden cursing as she makes her escape for the umpteenth time. Eventually she will make a mistake, and I will run her to ground.

Meringue has a good line in blood-curdling screams, which she employed when caught, causing Svenson to rush to the rescue, see me and run away again. This was only slightly gratifying to my masculine ego, because I knew that Svenson was essentially a bit crap at being a cockerel. Having retired to a safe distance, he watched and cursed as I inserted Meringue into her saddle. She never cooperates, with the result that the process always takes longer than it needs to. It also generates a fair bit of unnecessary mutual stress as I try to get the bird's wings through the elasticated straps without yanking her feathers about too much, then wrestle with the saddle so that it sits evenly on her back. Finally, with Meringue looking dishevelled and more than a little put out, I lowered the chicken to the ground, let her go, and she ran away at high speed.

To celebrate this happy moment, Svenson rushed over to try to give her the benefit of his private parts. Meringue resisted and did her best to run away from him. Embarrassment is not a feature of chicken social interaction, so Svenson gave chase, squashed her into the ground and engaged in some energetic, inconclusive backside waggling, the saddle now preventing his spurs from digging into her flesh.

It can have other disadvantages, though. After a particularly athletic, if unsuccessful, coming together (which involved a flying leap, a cockerel doing the splits and some issues over his braking distances) the bird more or less slid over the top of his intended and ended up in a heap in the grass, rucking up and pulling the chicken saddle over Meringue's head. This made her look as if she was wearing a large, soiled, Amish bonnet.

Having a big, looming thing over her head was alarming for Meringue, who did her best to get away from it by walking backwards, as if this would allow her to reverse from beneath it. Unfortunately the saddle came too, which caused the bird to try shaking it off by going backwards more quickly, scattering the other hens as she went in a series of figures of eight.

This did not make her easy to catch so that I could sort out the saddle; approaching from the rear was almost impossible since she was zigzagging dementedly. I was forced to go for a frontal assault, which of course meant she could see me coming. Since she was already trying to resist the saddle, Meringue quickly demonstrated her multi-tasking skills by resisting being caught as well. She might not have had eyes at the 'pencil sharpener' end of her body, but the bird proved remarkably quick and ever more stressed as I crouched, lunged and parried at her retreating form.

Eventually she became wedged into a bush and I was able to yank the saddle back into position and release the (by now) hysterical chicken, who immediately ran off to join Svenson, with predictable consequences. Since the idea of kitting all our flock with saddles during the summer months didn't appeal (although I have heard of someone who allegedly did this with patterned, home-made offerings for reasons of fashion rather than practicality), I decided it was time to do something about Svenson's spurs.

*****

It is possible to clip a cockerel's spurs, and we have a rather nasty-looking implement resembling a pair of pliers which does just that. Care has to be taken when using it, and with what it leaves behind on the cockerel. The sharp, end pieces of the spur are almost opaque, and it's important not to cut too deeply to avoid copious bleeding, if not obvious discomfort. The other thing is that you're left with something that's still rough edged and uncomfortable, and poor Svenson has become used to being held at an undignified angle whilst I manicure his just-truncated spurs with an old emery board of my wife's. Even this only worked up to a (lesser) point. Added to this, being handled at all caused the bird no end of stress. He would become rigid with tension, fanning his out neck feathers to express his displeasure, and I genuinely worried for his health.

Years before, we'd owned a particularly highly-strung cockerel called Zorro, who needed a trip to the vet. He found the whole process so alarming that he collapsed from shock and died soon afterwards. So I wanted something involving less cockerel interference that would also keep Svenson's ladies unsliced. The solution came in the unlikely form of a rusty old metal clothes horse we were about to throw away. The struts sticking up at the top were protected by little rubber caps in a violent shade of orange, designed to be thoroughly visible so as to discourage people from poking themselves in the eye. If they were silly enough to do this they would encounter a little rubber dome rather than something shaped like the end of a welding rod. The caps would still hurt, but were unlikely to blind.

I found they could be easily prised off, and looking at their undersides speculated how well they would fit over a male chicken's spur, and how they could be persuaded to stay there afterwards. High-impact glue seemed a good medium for attachment, as it was bloody difficult to unglue once it had set, but was the sort of thing beloved of people faced with carrying out emergency surgery on horribly gashed patients for whom there was no time to go to A&E.

With this lovely thought in mind I stole into the garden with a tube of uber-adhesive, and a pair of orange rubber clothes horse caps. Svenson saw me coming and made himself scarce. After about ten minutes of thrashing around bushes where the bird had taken refuge, then failing to catch him as he broke cover and sprinted away, I went to the shed and extracted an ex-fishing net. This had once been my Dad's and was getting tired, so he'd upgraded to something less weary, and gave me the old one as an ideal chicken snarer. It was large, a bit clumsy with its very long telescopic handle, and rather floppy, but otherwise looked ideal for the job.

As Svenson shot past me I lowered the net over his head and was gratified when the bird jumped upwards and more or less ran into it. I'd seen a professional chicken breeder do much the same. He said hens responded like this, apparently not realizing that it was a short cut to capture. So it was with Svenson, who was now somewhat trussed up.

'Sorry, mate,' I said to the bird through the thrashing net, aware as I did so that a single yellow eye was giving me a very dirty look from beneath its webbing. 'You won't enjoy the next bit, but everyone else will be pleased.' I edged towards the writhing form and prepared to extract the cockerel, something which resulted in a final burst of struggling.

Suddenly the struggling stopped, and the net was still. Had I killed another cockerel? No, but I realized that the net was empty and a slightly roughed-up chicken was sprinting triumphantly towards the other end of the garden. As far as Svenson was concerned, resistance hadn't been useless. Instead it had caused a seam between the netting and hooped frame to give way, creating a convenient, cockerel-sized hole through which he had escaped. I repaired the net with string, and ten minutes later cornered, and finally captured, the bird who'd hidden behind the summerhouse.

Clasping him under one arm and with a hand wrapped round his legs, his head pointing towards the ground and rouged, feathery bottom up in the air, I got to work on his spurs with the clippers and emery board, hoping he wouldn't express himself on my clothes.

Clipping and filing out of the way, I reached for the tube of glue. I found that unscrewing the lid and holding the cockerel required a certain amount of dexterity, but finally managed to squirt some glue inside a rubber cap then squeezed this over a blunted, and relatively benign, spur, holding it in place with a thumb and forefinger until the glue began working. Having repeated the process with the other spur, Svenson and I made a brief foray to the house where I picked up a tin of sweetcorn. Then we headed back to the hens' part of the garden. By now Svenson appeared resigned to the indignity of what was happening to him, and had relaxed a little, but even so, once his feet touched the ground he sprinted away.

However, as soon as some sweetcorn was sprinkled onto the grass, and his harem had thundered towards it, greed got the better of him and he pitched in with everyone else. I knew that we'd have to reglue the caps at some point, as the spurs grew and sharpened and the glue weakened, but he was completely indifferent to the rubber caps, which seemed to twinkle rather prettily when he ran.

Having fed he was soon road testing them by jumping up and down on Bella. Once this was out of the way she shook herself irritably, a few feathers falling to the ground as she did so. She was highly irritated all over again when I caught her a few minutes later to see if there was any sign of damage to her back. Parting her soft, almost downy, under plumage revealed cut-free flesh.

'I think you're in for a summer of love,' I said to the wriggling bird.

FOOD FIGHT

Meanwhile the chickens had rediscovered the delights of the fruit cage (or, more specifically, how to get into it).

I'd built this for Jane, and would like to offer a little DIY advice at this point. Domestic copper piping is not the ideal material to make a frame over which plant-protecting netting can be slung. It's soft, so it stretches, bows and fractures. I used it because it was easy to work with, and clearly hadn't thought through what that actually meant. The end result had a Tower of Pisa lean, and during the winter it would collapse if the wind blew too hard, or a few snowflakes landed on it. Keeping the saggy edifice going involved annual remedial work with bits of wood and a lot of cursing.

It was better than nothing, but not much. Starlings would find their way in, gorge themselves on my wife's fruit bushes, then fling themselves at the netting when they attempted to escape. We also noticed that something had been on a leak-chewing exercise, and shortly after that saw the first member of another year's rabbit plague helping itself, having found its way inside. There was also evidence of other foraging visitors that dug things up, and as the fruit cage became increasingly fruitless, I spent endless weekends trying to fix holes and gaps, only to discover new ones, mostly made by the rabbits.

One afternoon I found a raiding party of chickens, including Squawks, crashing about in the rhubarb. I watched as Nude launched herself into the air and yanked a fat raspberry from one of the canes. As I chased them out, I noticed something hiding among the vines in a corner where it was very hard to reach. 'Slasher. Out!'

The bird ignored me, and in the end I found a garden stake, and gave her a gentle prod. With that she took off like a rocket, revealing about thirty pale green eggs. She'd obviously been laying them here for weeks, within eyeballing distance of her previous pile (the one that Hoover had trashed earlier in the year). Once again, not being able to tell how old they were meant that the easiest thing to do was to chuck them away.

The wastefulness pained me, but it was better than food poisoning, so I bagged up and binned the eggs, then unsportingly put some bricks in Slasher's illicit nesting site, and spent another hour trying to make the fruit cage properly hen-proof. This slowed down the decimation of our small crops, and caused the hens to wander up and down, failing to find a way back in, something that almost made me feel bad for about a nanosecond, until I thought about all the food they did get. Gratifyingly, Slasher began laying her eggs in the henhouse, and I did wonder if that meant the problem was finally licked.

Next on the agenda was another round of bunnycide. By now the builders were three-quarters of the way through putting up the extension's outer shell, and one afternoon the Bricklayer had cheerfully called down to me from his scaffolding eerie that 'Your garden's crawling with bloody rabbits, isn't it?'

'You're welcome to them,' I'd said.

'They taste good,' said Number One Builder. 'We've still got some of the ones I shot in the freezer, so we haven't got room for any more.'

The Bricklayer looked thoughtful. 'My Mum makes good rabbit curry.'

I asked if he had a gun.

'Oh yes,' he said.

The next time I saw him I noticed an oil cloth containing an air gun among his tools, and retreated to my office, where I tried to ignore the soft 'thwack' noises made by the gun at intervals during the day. After he'd gone home I had yet another attempt at seeking out and plugging the holes his victims had made under our fences, knowing that after a respite they'd be replaced by more, and our garden would once again turn into a furry killing field.

BEAK ENCOUNTERS

Although we were the instigators of removing creatures from the garden, our habit of introducing new ones continued, and there was a sense of expectation when Jane and I went to let the ducks out for the first time. Their presence would inevitably change the feel of the place. We've wondered sometimes what it must be like for the new arrivals, having been uprooted from familiar surroundings and finding themselves in a completely transformed world. It happens to most of us at some point in our lives, but having language and the ability to think usually gives a handle on events and a way to rationalize why they've happened. When you're guided entirely by a mix of instinct and feelings that are as basic as *'I'm hungry/horny/happy/scared'* and so on how do you deal with absolute change?

So Jane and I try to be as gentle as possible with new arrivals to the garden, and as we got everything ready for the ducks' official introduction we tiptoed round their enclosure so as not to frighten them too much. Despite this they knew we were there, and began quacking loudly. The chickens were already up and in the garden, congregating nearby like rubberneckers at a road accident, waiting to see what would happen next. After Jane had filled a dog bowl with water and tipped in some duck food, I climbed into the area we'd fenced off for the ducks, fiddled with the catches of the 'Dalek', put down its little drawbridge and opened the doors.

We'd expected the ducks to stay inside until their unfamiliar human kidnapper had gone a safe distance away. Not so, and I'd barely unpenned them and straightened up when they charged into the sunlight. Unlike the previous evening, the chickens were transfixed. Fascinated by their new neighbours Meringue, Ann, Bella & Co. huddled together and stared. There was a different vibe to their interest. When we'd introduced new hens, kept in separate runs so that the existing birds could see – but not get at – them, the atmosphere had been distinctly unfriendly, with the old hands staring out the newcomers. This wasn't the same at all. The ducks weren't part of the chicken's flock, or potentially part of it, nor were they threatening. Instinctively, the hens knew what the dynamic was. Meanwhile the ducks knew what they wanted. Feeding.

We'd been soft-pedalling their introduction, if not to save their feelings then at least to reduce their stress. Almost at once it was obvious that we needn't have bothered; they weren't in the least bit worried. Having failed to notice the bowl of food, and deciding that I was the most likely source of something to eat, they rushed round in circles for a little then came stomping up and made sustained eye contact, looking at the ground, then up at me. They might not have had language skills but, just like the hens, the message was clear: *'Where is it then?'*

'Do we have any bread?' I asked.

'It's in the house,' said my wife, and headed for the kitchen, returning quickly with a bag of something wholemeal. I broke off a couple of bits and gently proffered them to the ducks.

The female skittered up, yanked her prize from my fingers and retired to ingest it with a manic energy. Her partner was more reticent, running up to me then running away again. It was obvious that he was in an agony of greed. He really, really wanted the bread, but couldn't quite bring himself to take it. Mrs Duck was braver. She brushed past him and went for seconds. Jane passed some more bread. By now the male was beside himself, and began making a strange, forward-rolling motion with his neck, so that his yellow beak almost touched the ground. He kept this up for a bit, stopped and looked me in the eye again. It dawned on me that – to use a hoary old Hollywood cliché – this duck was trying to tell me something. And that something was *'Put the bloody food on the ground, fool.'*

So I did what I was told. Mr Duck hoovered up the bread and retreated a few steps to gulp it down. As all this was going on his amour had discovered the bowl of Floating Discs and was getting it down her neck as fast as she could. I clambered over the fence and handed Jane the bread packet. Almost at once both ducks stared at her and began the neck-rolling thing.

'Go on. Bread here. Get on with it.'

'Well,' said Jane. 'They're not shy, are they?'

They certainly weren't. By comparison new chickens were thoroughly stand-offish. These two were already treating the garden as if they owned it. As is often the nature of chickens, ours had briefly been fascinated by the new arrivals, but just as quickly lost interest and had pushed off as things had become interesting. They reconvened at the far end of the garden, which in the morning was its warmest point. They milled about in a suntrap, a mix of whites, golds and green-tinged blues.

Having demolished their breakfast in very short order, the ducks noticed them for the first time, scuttled up to the fencing that kept everyone apart and gave the chickens their complete attention. Then they began running up and down, looking for a way to get out. The woman we'd bought them from had said that they had grown up with chickens, and were quite happy around them, and it quickly became obvious that they wanted to make the acquaintance of ours. Would this be a good idea? We'd always introduced new hens to the flock over a period of weeks. The ducks had only arrived the day before, but they seemed keen.

'You can catch them if there's trouble,' said Jane as I peeled back the fencing, and the ducks cannoned past us and hurtled towards the hens.

Words such as 'majestic', 'elegant' and 'athletic' can all be applied to some animals. They do not include Indian Runner ducks. Try 'daft', 'ungainly' and 'demented'. These suit a fast-moving Indian Runner, head stretched forward and wobbling from side to side, posture like a seventies' grandmother in a winter coat running for a bus. Quacking loudly, they shot into the middle of the knot of chickens, contentedly sunning themselves.

We half expected trouble. After all, this was the hens' territory, and having it invaded by this clumsy duo was unlikely to go down well. The hens were certainly surprised. There was a lot of rushing about, and a good deal of noise, but there was something about everyone's body language that suggested inter-species violence wasn't on the cards. We spent the next ten minutes watching and waiting to see if anything would kick off. When it didn't, and the hens' behaviour changed from surprise to indifference, we left them to it. We only realized later that we should have stuck to the avian apartheid for a couple of weeks to make sure that the ducks hadn't brought any chicken-harming diseases with them, but by then it was too late.

An hour later we sneaked back to see what was going on, and found the ducks wandering about in one part of the garden, and the hens in another, as if they'd known each other for years. Every so often the ducks would appear to remember the chickens, and hurtle across the lawn to join them. When they arrived, their presence barely seemed to register with the hens, who eventually wandered away from their uninvited guests, probing the lawn for treats as they went. As we watched them, I began thinking of suitable names for the ducks, and suggested one for the drake.

'No,' said Jane. 'We're not calling him that.'

'What's wrong with "Bombay"?'

'I think you're a very nasty person.'

'We could call her "Crispy".'

Jane gave me an acid look.

'We're not naming them after takeaway meals. Poor things.'

As an alternative I suggested calling the birds 'Number 6' and 'Number 27'.

'No,' said Jane.

So Bombay and Crispy it was.

OUT! OUT! OUT!

A few days later, with the duck/chicken assimilation apparently going well, we turned our attention to the doves. The time seemed right to let them out as well.

The infant dove was still a tiny pink blob, and had become spectacularly ugly. It looked like a creased, executive stress ball with spiky feather spines, a pair of very round eyes and a wobbly beak that seemed too big for its face, but its parents appeared devoted. If nothing else this would, we reckoned, keep these three in our garden even if all the other adult birds decamped in the way our first flock had. So one balmy evening after the builders had gone home we opened the dove prison doors, stood back and waited to see what happened.

For about ten minutes everybody stayed indoors, apparently completely institutionalized by their weeks of incarceration, until a white head tentatively poked out of a now open doorway. Eventually the head's owner launched itself unsteadily into the air and weaved drunkenly to the summerhouse roof, where it landed awkwardly and blinked at its surroundings. Soon some of the other adults followed suit, before returning to the dovecote to engage in a series of nest invasions and vicious skirmishes.

During the Industrial Revolution recently urbanized families apparently liked nothing better than to make their own entertainment by engaging in street wars, where the occupants of one row of terraces would swoop on another for an amusing evening of wanton destruction. So it was with the doves, and all we could do was to wait for the shriekings and wrenchings to stop. Somehow the infant dove survived several trampling episodes, and everyone else, having decided that domestic violence was too much like hard work, flopped onto the lawn to recover. Being white, they stood out, which of course made them interesting to the chickens. Meringue in particular felt their presence was an outrage. She stalked up behind one of the prone doves, vigorously jumped up and down on it and yanked out a beakful of feathers before her victim wriggled free and flew unsteadily onto the summer-house roof. It stood and peevishly preened itself, leaving Meringue to stomp off triumphantly in search of another victim. Perhaps the dove thought it was being beaten up by a distant older cousin with a weight problem. At least it had the sense to fly out of reach. Most of the others appeared to have forgotten what their wings were for and half flew, half ran to escape the marauding hen, who was clearly having a good time chasing them off, and was soon joined by some of the others.

In the end Jane and I rounded up the chickens and confined them to a run, from where they stared at the still-lawn-prone doves with ill-disguised enmity. Although our second flock of doves were feral, they also appeared to be completely useless at life as wild birds (which was a surprise, given that's what they'd been when Farmer Cuckoo Spit rounded them up a few weeks before).

As it grew dark the pair that had become parents managed to fly back to the dovecote and their now rather neglected baby. One of the other pairs did the same too. This was in mid-July, after a baking summer's day; the ground was still giving off heat and there was plenty of light in the sky well into the evening, which was why I quite often didn't lock the hens into the hen-house until late.

As I made my way there across ground which had taken on the consistency of rock thanks to a lack of rain, and guided by light from a full moon and a torch beam, I became aware of a pair of pale shapes in the middle of the lawn. Two of the doves had decided to roost on the ground, in about the most exposed spot they could find. Given that these were birds that could be jumped in broad daylight by a lumbering chicken, and didn't seem to find the presence of a bloke with a torch a cause for the least concern, I did not fancy their chances if a fox visited the garden. (Actually I didn't rate their ability to outwit a hedgehog with a limp.)

Back in the house I discussed tactics with my wife. We decided that they couldn't stay where they were, and would need to be helped back into the dovecote, probably best achieved by approaching them in a pincer movement from opposite ends of the garden and dropping a cardboard box on top of them.

'They're so thick!' I hissed as we tiptoed to our starting positions. Then we moved in on the birds, who seemed completely disinterested in us, until we got about three feet away, when they decided to run off. Flying, which would have made them impossible to catch, was clearly too much like hard work. Neither of them could be bothered to run very far, which on one level was good, but it was dark, the ground was uneven, and they were still little sods to catch. We spent what seemed like an eternity running, tripping, lunging and dropping the cardboard box over where the doves were, but always a split second too late. Every time a small white shape would scuttle out of the way at the strategic moment, then stop about five feet away to get its breath back. The night air carried the sounds of two middle-aged people running about in the dark, slamming an upturned cardboard box onto a lawn as they muttered 'Over here!', 'Nearly,' and 'Oi! Come back!'

Eventually Jane plonked the box down rather like a rugby player scoring a try and shouted 'Got one!'

Lying on my stomach I wormed my hand underneath the box, trying to keep the gap between box and ground as small as possible. I felt around until I encountered something warm, rounded and feathery cringing in the furthest corner. Wrapping my hand round it as gently as I could, I could feel the dove breathing fast, its chest expanding and shrinking. Lifting it off the ground I felt its hard, dry little feet and claws brush against my hand as it kicked uselessly. Now I had the task of getting up the ladder in the dark, holding a dove in one hand, not falling off, and posting the bird into an unoccupied nest. Somehow I managed this, and then it was back to the chase.

Now you would think that the other dove, having seen what was going on, would have done the sensible thing and made itself scarce. Oh no. It sat on the lawn and waited for hostilities to begin again. Eventually it too vanished under the box and I made my way up the ladder with another unwilling fellow traveller, before Jane and I went back to the house, having decided that something alcoholic was needed to calm our nerves.

The following day the doves decided that the summerhouse roof was safer than the lawn, and watched as I went up and down the ladder to furnish them with food and water. All the doves that is, except one of the parents, which sat on its nest, puffed up its feathers and swore as a big fat human face blotted out its view. The others waited until I was safely back on the ground before heading out and eating.

Soon a gaggle of hens congregated at the base of the dovecote and skirmished over the extra pickings that fell from it. If it wasn't possible to beat up the doves in person, eating their food was certainly a good consolation prize.

Things quickly settled into a bit of a rhythm, and I began to think that the low-maintenance aspect of keeping doves might not be a fantasy after all. They soon decided that they could be bothered to fly, and although they still found the ground an agreeable place to lounge around on bakingly hot summer days, they became more adept at interpreting the sound of thudding feet as a chicken with ill intent bearing down on them, and realizing it was time to get out of the way.

We'd been given a slightly twee bird-bath, made of plastic distressed to look like dark-green marble. It came in three pieces: base, column, top bit. Each section had threaded ends, and assembling it should have taken less than a minute, but in the course of first putting it together I'd over-tightened the bottom part. There had been a sickening 'crack' as the base snapped at the thread end. Fortunately the base was hollow, so I jammed an oversized piece of wood inside, ramming it into place. The end result was pretty stable, if now lop-sided, but our enthusiasm for having this particular piece of garden furniture near the house had waned somewhat.

However, at the far end of the garden it made an ideal dove water receptacle. They could use it without being mugged by the chickens.

HOME IMPROVEMENTS

Thanks to the builders the house had changed shape. There was something weirdly balletic about the way they worked, making the same movements over and over again as the extension grew. I'd watch Knocker and Male Bimbo alternate with a concrete mixer and wheel-barrow, moving backwards and forwards supplying cement or bricks to the Bricklayer, who in turn was helped by the Boss and Number One Builder. They moved as a unit, in a way that reminded me a little of the domestic birds who occupied the rest of the garden.

I also began to appreciate the artistry and craft that underpinned what looked like repetitive manual labour. I'd watch the Bricklayer using little more than a spirit level, lengths of string and his eye to lay row upon row of perfectly aligned brickwork.

'That's perfect,' said John, one of our village's elder statesmen. A short, white-haired character in his late seventies, whose family has been here long enough for one of his ancestors to have nearly booked a passage on the *Titanic* when they'd decided to emigrate, John is one of those permanently busy pensioners who nevertheless has time to chat. He's the sort of person you're always pleased to see, but he does live in that older person's universe where time and deadlines are meaningless.

His habit of stopping you doing what you were doing to talk about something he'd seen, or tell a dirty joke, made him both refreshing company and a bit of a challenge when you were in a hurry. Both I and the builders had discovered, invariably, that he was far more interesting than the things we had to do.

He'd appeared one sunny afternoon to appreciate their work, which had of course stopped as a result. I'd discovered this inactivity, and joined in by making tea for everyone. In between discoursing about life and the universe John ran a practised eye over a recently laid flank of bricks.

'Next time you're out you look at a wall or a new house,' he said to me when the Bricklayer wasn't in earshot. 'Bet you'll see bricks sticking out at different angles, but this is absolutely spot on. Not one out of place.' John squashed the side of his face against the wall and squinted along it. 'This boy's good. You didn't have any idea how lucky you were to have him, did you?'

Although I'd missed this, it was obvious once John had pointed it out. It also made me think about the henhouse, which really needed some professional care too. It wasn't going to get it, since I was the person who infrequently maintained it, but feeling curiously inspired I decided to give it some amateur attention, and to at least try doing this properly.

'Are you feeling all right?' asked my wife when I told her, especially when she discovered that the work involved a trip to a DIY megastore, one of my least favourite places. Still, a small, brief *frisson* of excitement had passed through me on discovering it had a special offer on paving slabs, which I wanted to use to replace the henhouse's wooden floor.

'I think this proves you're now officially middle-aged,' said Jane, as I set off.

By the time I started work on the henhouse, the builders had finished the extension walls and built a skeleton of wooden roof joists, so our home had both changed shape and looked all of a piece. I'd noticed a certain masculine pecking order between them, with the usual banter and easy-going verbal put-downs, which seemed to be dished out more by the Boss and Number One Builder, and tended to be directed at the younger Male Bimbo Builder, who at lunchtimes had become enslaved to our dog's Scotch egg addiction. The way he caved in to Hoover's professional winsomeness was the cause of much amusement. There was some joshing aimed at the Bricklayer, who was a similar age to the older pair, so there was more equality to this.

I noticed that although the remarks were sometimes near the knuckle, they never crossed the line into bullying or active cruelty, because they were underpinned by friendships, and were part of the lubricant of the builders' working relationships rather than the grit which jammed them up. (I also noticed that nobody took the piss out of Knocker.)

I thought about this as I jemmied up the henhouse floor, because it reminded me of past school and work relationships I'd experienced. And as a gaggle of chickens watched me rip their home to pieces, I realized once again that it was even a little like the way the birds worked together, but without the free market-style *'I'll have that!'* approach to food or the casual violence.

Once I'd got the old floor up I discovered two things: a network of little tunnels indicating that we'd had mice, and the fact that the henhouse was now distinctly droopy and the main door no longer shut properly. Things weren't going to plan, but they were conforming to a familiar pattern of a sudden rush of enthusiasm, a lack of planning and a lot of *ad hoc* remedial work to produce something that was now mildly lopsided and had a slightly uneven stone floor. By the time I'd finished I was hot, grubby and had completely run out of enthusiasm, but at least the henhouse was still standing, would be easier to clean out and drier too, as the old floor had been made of a blotting-paper-like chipboard. It would also be rodent-proof because the paving-slab floor was sitting on a bed of gravel. Even the most determined rat wouldn't be able to tunnel through it.

'What are you going to work at next?' asked Jane before I headed for the shower.

'How about writing a best-selling novel?' I scowled. 'Then we can pay someone else to sort these things properly.'

PARENTING ISSUES

The baby dove remained lumpen and pretty immobile, but it was keeping its parents busy with constant demands for food. In the henhouse maternal feelings were stirring too. Ann Summers, one of our more venerable hens, was showing signs of going broody, which given her advancing years was a surprise. Our initial reaction was that this should be discouraged, since there was enough going on with the ducks, doves and builders, and we felt that more chickens was something we could do without, especially if she hatched out cockerels.

Ann had proved her parenting skills with Ulrika, Svenson, Bella and another bird we'd christened Bonnie. These last two had been named after our two village pubs. There are various ways of acquiring chickens, but these can generally be split into two categories: hatch them yourself, or buy them from reputable breeders. Getting them from a man in the pub hardly fits the bill, but it was through this slightly dodgy-sounding medium that Bonny and Bella appeared a previous summer when our lives were still dove- and duck-less. They arrived as eggs, and the person who supplied them and a large tray of other eggs did not fit the cliché of a bloke offloading pirate DVDs in the snug of a back-street boozer. One of the pubs has a rather lovely garden, a suntrap in front of the building, which is opposite the church and sits on the nearest thing to a main road the place has.

It's great for people watching, and in the summer it's a bit of a village meeting place. Hoover loves it because the landlord, a well-built, properly publican-shaped man in his fifties, likes dogs and keeps a jar of Gravy Bones on the bar, and one of his regular barmaids is rather besotted with our dog, and will offer him bits of ham.

The place is also a melting pot of village life, and a repository for what might be described as 'local colour'. So you can find the doctor rubbing shoulders with Oscar, a genial, round man in his eighties, who used to drive giant diggers, was born in the village, and has the build and face of Tweedle Dee from *Alice in Wonderland* (although, unlike Tweedle Dee, Oscar has a penchant for bobble hats).

You'll see men with beards, some of whom are unashamed Morris dancers, chatting with a bloke who writes movie scripts, and the retired, cravat-wearing banker and his effortlessly elegant wife. On the surface both appear very English and conventional, but possess a shared dry sense of humour and that worldly twinkle of people who've had interesting lives. They're great friends with a stick-thin, entertainingly camp retired dentist, who has a waspish turn of phrase and an observational eye. They'd invited us to a party where this man shimmied from guest to guest topping up their glasses. We'd fallen into conversation about birdwatching with a formidably patrician old lady in county shoes and a floral print dress. The dentist appeared with a bottle of something bubbly, and the promise of further refills.

'When you're ready,' he said, in a deadpan stage whisper, 'just snap your fingers and I'll come.'

As this *double entendre* passed over the head of our companion, the ex-dentist melted back into the crowd.

So as a consequence of all this local colour, the characters and conversation at the pub are always pleasingly mixed. During the summer months its garden is alive with families, and a gaggle of cheerful thirty- and forty-something couples, their friends, parents, children and dogs. There's a sense that a lot of these people grew up together, and will probably grow old together too. There's an easy, almost hippyish feel to their gatherings, which are good-natured and seem to go on for a long time.

We met the man with the chicken eggs thanks to his dog, a young lurcher with legs and fur that stuck out at wild angles, and a permanently inquisitive nose that wanted to make Hoover's intimate acquaintance. The lurcher's lead had become detached from the pub table to which she'd been secured, and the dog had bounded over and stuffed her face into one of Hoover's more personal crevices. Judging by the way he flattened his ears and showed us the whites of his eyes, he found this a mixed blessing.

'Well, that's what it's like when you do that to the cat,' muttered Jane.

'Sorry, sorry,' said the lurcher's owner, a stubble-headed, round-faced man of indeterminate middle years.

I can't remember how the conversation came round to chicken keeping, but our new friend, who said his name was Jamie, ended up sitting at our table talking about his birds. 'I breed them at my Mum's,' he said. 'I've got Buffs, Araucanas, all sorts.'

This caught Jane's interest. Araucanas lay blue/green eggs, and we'd discussed the idea of taking on a couple. We dithered pleasurably over the idea as Jamie chattered on amiably.

'I've got fertile eggs if you want those,' he said.

This got my attention, because Ann Summers had gone broody that year too. We weren't especially keen on hatching crossbreeds from our existing flock, but the idea of rearing our own Araucanas rather appealed. Jane, not unreasonably, suggested that we might hatch out nothing but cockerels, which would be hell to rehome, but despite this, the more I thought about it, the more the idea appealed.

'It wouldn't hurt, would it?' I said to Jane.

She laughed. 'If they're all boys it's you who'll have to find them new homes.'

'Well,' said Jamie, standing up and hauling on the lurcher's lead. 'I'm here most evenings after six. You come along on Tuesday and I'll have some eggs for you.'

We asked him how much he would want for them, and he shrugged, blew a raspberry and said that it didn't matter.

A couple of days later I put in an appearance at the pub, as arranged. Jamie had wasted no time in getting stuck in to the local brew, and was smiling beatifically as he pointed to a huge egg tray filled with about twenty eggs of various sizes and colours.

'I thought you'd like a selection,' he explained. 'The green ones are the Araucana eggs, I think, but you can just stick some under your broody and see what hatches out.'

Chicken lucky dip wasn't quite what we'd had in mind, but what the hell?

'I really should give you some money for them,' I said. 'Don't be daft.'

'Can I at least buy you a pint?' I said. My new friend's resolve wavered slightly, so pressing home my advantage, I found out what he was drinking, and handed over the price of another pint of the same.

We live about half a mile from the pub and I'd walked. Making the journey in reverse with a large, flexible container of fragile eggs took rather longer. At one point it sagged dangerously in the middle and some of its precious contents dislodged, forcing me to lower everything gently to the pavement, carefully replace the eggs then rise slowly and continue to totter homeward.

Ann was not pleased to see me, because the hot, hormonal fury that goes with being a broody chicken had gripped her. She sat in the nest box, sizzling with rage, neck feathers fluffed, her wings covering other chicken's eggs. She was warming these with a near naked underside, since most of the feathers there had dropped out, the better to superheat the next generation of chickens.

She was ensconced in a gloomy corner and very reluctant to move. At mealtimes she more or less had to be crowbarred away from anything that resembled an egg or she wouldn't bother to feed. After being manhandled she would fling herself into the garden, shrieking dementedly, barging into the usual bunfight of feeding chickens, grumbling and pecking at the other birds as she went. Food would be consumed ravenously then, still cursing, she would stump up the drinker, imbibe copiously then vanish back into the henhouse, once again hunkering down onto the eggs.

We had bought a rabbit-hutch arrangement with a wired-off run for the purposes of breeding chicks, and this was where I planned to put her and Jamie's eggs. The wire mesh was narrow and small, rather than regular chicken wire, which we knew from personal, unhappy experience was large enough for a very young chick to squeeze through. One of our very first chickens was a beautiful but rather butch bird with glossy black feathers, whom we'd christened Elvis. The butchness manifested itself under duress, when Elvis would utter strangulated semi-crowings. She would also lose vast quantities of feathers whenever she moulted, changing from being elegant to a piebald mix of pink skin and feather quills, so that she looked as if someone had started to pluck her but given up halfway through.

Her feminine side was apparent in her willingness to lay eggs, and a habit of growing broody every summer. For about three years on the trot we sat her on clutches of eggs, but for the first couple none was fertile. Then, when she was getting quite old, her luck changed and she became the parent of a tiny black-and-yellow chick. Elvis was a solicitous parent, and her baby emerged as an active, healthy little bird, but we had corralled them in a run with adult-sized chicken wire. The tiny creature squeezed through, and couldn't work out how to get back to its mother.

Jane came back to find it cold and weak on the wrong side of the wire, Elvis beside herself on the other. I'd been working in London when Jane rang, frantically asking what she should do. We decided to give it a little water with an eyedropper, but distressingly, the tiny creature died in Jane's hand as she was speaking to me. This was the last time that Elvis went broody.

Having learned the hard way that accidental human negligence can have tragic consequences, we were determined not to make the same mistake again, and the birthing area earmarked for Ann and her eggs was altogether more secure. (Jamie had given us so many that I confess several ended up in a cake. 'Here's to infanticide,' said Jane afterwards, cutting a slice of something involving sponge.)

We selected eight that we thought should find their way under Ann, with six of these being the blue/green Araucana eggs.

'God knows what sort of birds laid the other two,' I said. 'Let's stick them under Ann and see what hatches out.'

They all looked very pretty on a clean bed of straw in the compact and *bijou* ex-rabbit house, and I decided to move their potential stepmother into it at dusk, so that she could be shut up for the night with her new eggs, and get used to the change of surroundings. Hauling Ann from her existing clutch proved mutually stressful. All the hens had gone to bed when I lifted the lid of the nest box and fastened my hands round her. She let out a scream that would have been a credit to Fay Wray in the original *King Kong* movie, and this caused everybody else to have hysterics. As I lifted her from the nest there was a lot of running about in the semi-darkness, accompanied by cluckings and crowing.

Ann was rigid and all but vibrating with tension as I carried her to the birthing box, stuffed her inside and shut the door on another piercing scream of terror and fury. The following morning she seemed to have recovered her poise and bonded with her new eggs, to which she was now firmly attached. I'd filled a small drinker for her and tipped a handful of food into a bowl. She could see these through the open doorway of her nest box, but showed no inclination to make use of them.

Another, rather unlovely feature of being broody is that the bird will spend hours on end not going to the bog. Chicken guano is a mix of urine and faeces, and the idea of this stuff curdling inside a bird that is neither feeding nor moving very much isn't good. So at meal-times I help things along by shoving the hen off her nest. I'd just done this when I noticed something different, and it took me a little while to work out what it was. Ann had been sitting on nine eggs, which was odd, given that I'd only put eight in the nest. Then I realized what she must have done. During the previous night's fraught move she'd clutched one of her old eggs and held onto it as I'd carried her away from the nest, something which made me feel a bit of a heel for not noticing. It was also a reminder of the life force that goes on unnoticed most of the time. The genetic imperative which drives the big, fundamental things in life can also be found in an English garden's henhouse. It seemed obvious that Ann's special egg more than deserved its place in the hatching clutch, and I left it alone.

Ann crapped, ate and drank peevishly, then went back to her eggs, which she gathered up beneath her wings.

Then she sat stock-still, eyeballing me with deep unfriendliness. Beating a hasty retreat, I wondered whether she would have held onto that egg had it been a fox rather than a man carrying her off, deciding the answer was obvious.

*****

It takes three weeks for a baby chick to gestate, and as the hatching day gets closer, mothers-to-be become ever less keen to leave their eggs. So it was with Ann, and with about four days to go before our clutch of pub eggs (and the one she'd brought with her) would be fully cooked, she had become incredibly protective of them. I'd extract her from the nest (being careful that she did not hang on to any of her eggs, in case one dropped and smashed), but as soon as I put her down near food and water she would rush back into the rabbit house and reattach herself.

'You'll starve to death,' I said. The bird had certainly lost weight.

Eventually I shut her out from her eggs for half an hour or so, during which time she would, eventually, eat and drink, although this was often prefaced by a sort of sit-down protest where she'd plant herself on the grass as if still brooding her eggs. To the accompaniment of much furious clucking, I'd chivvy Ann to her feet once more and shove her in the direction of the grain and water. Then I'd leave her alone to eat, more or less unobserved.

I say 'more or less' because the other chickens, rooting about outside the rabbit house and its run, knew something was going on, and often congregated round it. As Ann's confinement neared its conclusion, the other girls would be around more. I'd often find Brahms looking through the mesh into the gloomy recesses of the rabbit hutch, whilst Meringue clambered onto the roof and peered down.

Then I did something very stupid. One breakfast-time Ann was especially recalcitrant, so I shut her out and went back to the house, leaving her swearing over a tin bowl of layers pellets.

Once inside I became distracted by work, and as the day progressed felt irritated that it was keeping me in the house when outside the sun blazed and the weather was superb. By mid-afternoon I began to feel a rising sense of unease. I have what an uncle of mine described as 'a leaky memory'. To make things happen in our house a lot of stuff has to be written down, and I have to remember to look at what I've written. If this doesn't happen, some of the more inaccessible bits of my brain will occasionally shudder into life with non-specific warnings about forgotten tasks. The uneasy feeling is usually a precursor to this. So I knew I had forgotten something, but was not sure what. Then the image of Ann shut off from her eggs floated into my consciousness and I was running into the garden, hoping I wasn't too late.

The bird was almost beside herself as I reached the birthing box, and flung open the door. She stopped scuttling up and down the run and rushed to her eggs. There was nothing more to be done. The weather had been very warm and sultry and the eggs had been shut into a hot wooden box. If anything was brewing inside any of them this might have been enough to keep it alive, but since Ann had begun sitting on them for the previous nineteen days we'd know very soon.

It was a weekend when the eggs were due to hatch, and during the day I left well alone. By the evening anticipation and curiosity had got the better of Jane and I, so we trooped outside and gingerly lifted the lid of the birthing box.

Ann greeted us with ruffled feathers and a loud screech. We couldn't hear anything else, no cheeping or scrabbling noises. Then Jane noticed that the bottom of one of Ann's wings seemed to be moving. Leaning forward, I gently lifted the wing with an index finger. Ann screeched again, and something gold and fluffy wriggled into what remained of her underbelly feathers. I let the wing drop, and we left the bird to her own devices.

It turned out that Ann had produced two babies. One had emerged from an Araucana egg, the other from one of Jamie's indeterminate offerings. We were only slightly sad that the egg Ann brought with her hadn't been fertile, and after another forty-eight hours we disposed of this and the others that had failed to brew new birds. We were lucky in that both our new arrivals turned out to be girls, and we christened them Bonny and Bella in honour of our two village pubs, the Bonny Cravat and the Six Bells (where we'd met Jamie to collect them in egg form). Like people, some chickens have better parenting skills than others, but Ann was a natural mother, and her babies grew strong and feisty. They might not have a shared gene pool, but they seemed to have a similar strength of character.

Bella, the Araucana, lived up to her name. She was especially pretty, with ruffled, almost shiny, grey plumage. And she did lay green eggs. Bonny was altogether more stolid, with sensible white feathers contesting with a ruff of black, almost diamond-shaped, plumage at the base of her neck, a bit like a collar. We've never been sure what her antecedence is, and to be honest, don't care.

Being Ann's stepchildren apparently gave them both a head start in chicken society, and the pair assumed places fairly high up the pecking order. Sadly, Bella didn't last long, suddenly keeling over when still young, but Bonny and Ann just kept going, with the latter assuming the role of chief chicken. By the time the doves, ducks, rabbits and builders had arrived she'd long since given up egg laying, and had become a stately matriarch, hence our surprise when she went broody yet again.

This year we had nothing for her to hatch, and wondered vaguely if the strain of becoming a parent might be a bit much for such a venerable bird. I'd taken to regularly visiting the garden and hauling her off any eggs that had been laid, and taking them away. As the summer progressed this became a battle of wills, which Ann continued to win by skulking into the nest box even if there was nothing to sit on. Given that this was where almost everyone else was laying, I couldn't just shut it off during the day, something that would have eventually caused the bird to give up.

STASHER

'Has Slasher stopped laying?' asked Jane.

Once again the supply of her distinctive eggs had petered out, and since most of the other birds were all busy laying, we suspected that Slasher was too, but was again choosing to do this elsewhere. This time finding them proved easier. In a corner of the garden where the hens had turned a flower bed into a dust bath was a large, neglected shrub. At its base I found a familiar stash of green eggs.

She'd secreted about ten there. Yet again we'd have to bin them, but I decided that compromise might save the day. This time I left a single egg, marking it with a pencil, and was rewarded a day or so later by the arrival of a second green egg next to it. Slasher could keep laying here and they wouldn't be wasted.

'That's fooled you,' I said as she shot by, comb flopping up and down as she hurtled into the distance.

Slasher wasn't the only bird to lay green eggs. Crispy the duck had begun producing them too and they were so huge I didn't like to think of the work involved in getting them out. On most days I'd find one in the straw of the nest she and Bombay shared. However, there was little chance of her hatching anything from them. Indian Runners tend not to become broody, so anyone wanting to breed them needs an incubator or a surrogate parent in the form of another breed of duck, or even a chicken.

Certainly Crispy showed little interest in her eggs, but I did begin wondering whether their arrival had presaged a hormonal change in both ducks, who seemed a little more skittish than when they'd first come to us. They were friendly towards people, to the extent that they would eat out of our hands, but relations with the chickens – which had started so well – appeared to be getting strained. We'd taken to feeding the ducks and hens together, and initially there had been a surprising amount of tolerance; but latterly both ducks had started chasing the chickens off, lunging at them until Svenson intervened, at which point it would be the ducks that decamped.

Jane and I discussed what to do, and that's when she suggested digging a proper pond for them. 'If they're on the water they'll be too busy to hassle the chickens,' she reasoned. I thought this sounded plausible, but wondered where the pond might go. Jane had been thinking about this too.

'That pine tree,' she said, indicating a large, mature specimen about three-quarters of the way up the garden, and about fifty feet from the house. 'It probably has a big root ball, and if we got rid of it there would be a nice, pond-sized hole in the ground.'

Now I rather liked this tree, but Jane had spent years muttering that it blocked out the light and cast shadows on the garden, and I didn't like it that much. Reasoning that Knocker could probably tear it out of the ground with his bare hands, we asked the builders but they were unenthused, having enough work to do on our house as it was.

'Also, the diggers we used for the foundations were hired in, and they've gone. Getting them back would be expensive,' said Boss Builder, with a finality that translated as 'Thanks, but no thanks'. So we enquired in the village.

Where we live is pretty self-sufficient, partly because it isn't on the way to anywhere else, and that's one of the reasons we like it. As well as the shops and pubs, if a car needs fixing two garages are within walking distance, and electricians, carpenters and plumbers all live nearby. There's also Mr Lloyd, who puts up fences, landscapes gardens, takes down trees and digs ponds. A solid, avuncular man of indeterminate late middle age, Mr Lloyd worked with his two, softly spoken, gently hulking sons. All three have the easy-going confidence of people who are good at what they do.

Also, they lack the chippiness or faux servility some so-called tradespeople adopt when dealing with white-collar types like Jane and I, who know what we want – or believe we do – but have absolutely no idea how to go about getting it. And that's how we found ourselves in the garden with Mr Lloyd, looking up at the pine tree, discussing tactics for its demise.

'Got a woodburning stove?' he asked. We said no, but there were open fireplaces in the house.

'If you like we can chop that up and you can burn it. It'll save time taking it away.'

He agreed that yes, the hole left by the roots would be a good starting point for a duck pond, and that although the job would be fairly labour-intensive, it looked pretty straightforward. By the middle of summer the tree was gone and we had a large hole in the ground. Not long after that it was a hole filled with water. Since duck/chicken relations had continued to sour, we'd temporarily confined Bombay and Crispy to the other end of the garden, in their small fenced-off area with the Dalek and their faux tin bath, which they occasionally sat in, but otherwise treated with disdain.

When their pond was finally ready we grounded the chickens, who had spent the previous week wandering round the garden paying the imprisoned ducks occasional visits to wind them up. Hoping Bombay and Crispy would be delighted with their pond, we let them out and chivvied them towards it. They ran up to the pond, studied it, then turned and shot off in the opposite direction.

Again we herded. Again they approached, looked worried and peeled off in opposite directions, meeting up at the far end of the garden, at just about the furthest point from the pond that they could manage. A week later they still hadn't taken the plunge. Instead, they had begun to actively antagonize the hens.

'Perhaps we should fence them in with the pond. Then they won't be able to ignore it,' I muttered, thinking of the expense that had gone into sinking a large hole in the ground for their benefit, only for it to be completely ignored. Jane thought this might be a way forward, but also began musing about increasing the size of the flock 'now we have a proper pond'.

This gave me another idea. Ann was still thoroughly gripped by broodiness, despite my best efforts to discourage her, and Crispy was laying a fantastic quantity of eggs. Why not give Ann what she wanted, in the shape of a stab at older parenthood, but this time by getting her to sit on some Indian Runner eggs?

We were discussing this as we went to collect our new duckhouse, which belonged to Chris and Chris, the married couple who lived a few doors away and who'd briefly fostered Barry the cockerel. He'd lived there with his harem of Marans hens, and had apparently moved on just in time. Thanks to a passing fox, this birdhouse was empty, and its owners had concluded that they could live without the carnage of another attack, so had decided to give up their chickens.

'How much do you want for it?' we'd asked the male half of the partnership.

Chris shrugged. 'You can have it. Look, it's pretty old, and it's in the way. We were going to burn it.'

So we arrived at their house with our car, towing the small, tatty trailer we use to cart away the endless tangle of bramble and grass cuttings our garden always seems to produce. We perched the birdhouse on this, strapped it down as best we could, then crept very slowly the hundred or so yards back home. Chris, his teenaged son and one of his friends helped us pilot the weighty, bucking trailer across our uneven lawn, before we slipped and struggled to an area behind the pond, where we planned that the birdhouse should take up permanent residence. When neither it nor any of us fell into the water we were relieved and suitably grateful.

Having scrubbed out and disinfected the duckhouse I left it open to the elements so that it could dry out properly. Bombay and Crispy remained in the far corner of the garden, indicating that they had absolutely no intention of getting anywhere near this thing, at least not without some outside assistance.

PARENTING CLASSES

As the ducks continued to refuse to move into their new home, I discovered something had prevented the dove parents from getting into theirs to look after their lumpen baby. It was dusk and getting dark when I noticed that some of the wood I'd stuffed into the gap between their one-time prison box and the walls of the dovecote to stop the baby dove from falling out had fallen sideways, blocking the entrance to their nest area.

I cleared these out of the way and was met by the unblinking stare of the baby dove, whose parents were now nowhere to be seen. The days had been hot, but the nights had become unseasonably cold, and when I went to lock up the chickens at about 10.30 p.m. and shined my torch into the dovecote, its infant occupant was still very much on its own. Having been unable to get to it, the bird's parents had roosted elsewhere. Feeling bad that the expedient way I'd stuffed those bits of wood into the gaps had created the problem, and worried that the tiny creature might not survive the night, I went back to the house and told Jane what I'd found.

'It might be all right, but I don't want to risk it,' I said. 'We'll have to do something.'

'No,' said my wife, *'you'll* have to do something. I'm off to bed.'

Ten minutes later I manoeuvred the dog cage onto the tool shed, found some bricks, put these inside the cage, and placed an old electric tube heater on top of them. Designed to keep greenhouse plants warm without frazzling them, I thought it would be a reasonable source of heat for a baby bird. Then I found a very small cardboard box that had once contained a dog's flashing toy ball. The box was square and had tall sides. I put a little hay in this, snapped on a pair of latex gloves and went back to the dovecote.

The baby bird's unblinking eye again regarded me steadily in the beam of torchlight, as I scooped up the small creature, which was gratifyingly warm (and felt a little like a slightly squishy stress ball, but with stubble). I placed it inside the box and made my way back to the tool shed, praying that I wouldn't trip in the darkness. The box and its contents reached their destination unscathed. Its fragile cargo fitted neatly under the tube heater, which I was confident was close enough to keep the baby dove warm, but far enough away that the bird wouldn't get too hot.

I was less convinced that this was the right thing to do. I'd worn the gloves so the parents wouldn't detect my scent on their baby, and decide it wasn't their baby at all, but I wasn't certain how effective this would be. Then something rather sweet, if not necessarily good, happened. The baby stretched its neck upward and opened its wonky beak in a classic *'Feed me!'* stance. I couldn't tell how long it had been separated from its parents, and wondered about trying to feed it a mix of milk and breadcrumbs. The problem was that I had no idea what baby doves ate, and risked poisoning the one I was trying to save, so I decided against this.

'How's the patient?' asked Jane blearily, as I climbed into bed.

'Hungry,' I said.

'When do you want me to set the alarm?'

'4.30, I think.'

'Great.'

It seemed about five minutes later when I tottered out of bed, hauled on some clothes and another set of latex gloves and made for the tool shed, where I half expected to find a small, cold bundle. Instead I was greeted by a lot of animation and a familiar *'Feed me!'* stance.

We made our way back to the dovecote and up the ladder, where I scooped up my companion and put it back in the nest area. It was hardly worth going back to bed as I was heading to Goodwood House in Sussex to report on the Goodwood Festival of Speed, an old car event, where fabulously valuable classics – the sort of things that are worth more than your house and mine put together – are exhibited and raced. I spent the day at what *Private Eye* might have described as 'an agreeable stately home' in warm summer sunshine, surrounded by the automotive equivalent of the Crown Jewels, thinking about what was happening to a small, ugly baby bird in a Kentish garden.

One of my tasks was to interview Nick Mason, the Pink Floyd drummer, and long-time preserver and racer of some very beautiful, very valuable cars. A scrupulously polite and very accommodating interviewee, Mason is nevertheless well versed in dealing with journalists, having had a lifetime of being asked questions. His answers can sometimes be a little elliptical, especially if a question is lazily phrased or the questioner's mind is elsewhere. I don't think this is deliberate, or cussed, but I knew I would have to concentrate. Not long before I was due to spend a snatched five minutes with him, my pocket began to vibrate as my mobile phone went off.

'Jane' read the display. 'They're back,' she said simply. I had a blank moment. 'Who?'

'The doves. They're back with the baby. I haven't been to look, but they appeared about half an hour ago.'

The interview went well, and afterwards I had a very pleasant day being what one friend only slightly disparagingly described as 'a tyre sniffer'.

Back home and up the ladder I found parents and child at home and *en famille*, and decided that we'd had a narrow escape, and that things would now be fine. Of course just thinking this was a challenge to Fate, which was bound to go out of its way to prove me wrong. It waited until the following evening, when I made a nocturnal ladder check and found Junior on its own once again. Since it was now pitch-black outside, it was obvious the parents were out on the town somewhere or, if not, then at least roosting in a tree.

On went the gloves, tube heater and alarm. Once again I carried the tiny – although noticeably less tiny after just forty-eight hours – bird to the tool shed. The awful alarm shook me awake a few hours later, the process was reversed, and as dawn light made its way slowly across the sky, the parents returned. The next night I found the baby neglected again, but had a small brainwave, and looked for the parents in one of the dovecote's other pop holes. I found them next door to their infant, blissfully snuggled up together and not keen to be moved.

Extracting them resulted in much dove wriggling and muttering, but as soon as they were reunited with Junior, they fell upon it in a way that – had people been involved – could almost be described as 'a tearful reunion': the baby squeaking plaintively as its mother rushed forward, calling the dove equivalent of *'Darling! Darling! Darling!'*, then sitting on it firmly. The following night exactly the same thing happened, but this time with the added *frisson* that the parents had chosen a different part of the dovecote in which to roost.

Hauling them out I was relieved to discover that I had got the right birds, but working out that there would be trouble ahead if I chose the wrong ones, I came back the next morning with a red marker pen and daubed each parent's wing with it, something that was not well received. Perhaps this was why that evening they didn't show up at all, and I was again forced to extract their now even fatter baby, which spent a third night under the tube heater.

'They're latch-key parents,' said Jenny, our friend from Greenwich. 'Before you know it they'll start stabbing your chickens.'

Maybe it was this cheerful image that made me decide to impose the avian equivalent of a control order on the gormless adult birds when they duly reappeared. After doing the usual thing of roosting in the wrong nest site, so their baby could freeze to death in private, I shoved them back in with it – resulting in the now familiar besotted parent noises – and slammed down the door of their prison box. They were going to be grounded until Junior was ready to fledge.

RABBITING ON

The rabbits were back, boring fox-friendly holes under our fences (next to the ones I'd already plugged) and breeding copiously. Their babies looked cute as they gambolled over our lawn. Soon their parents broke into the veg patch for the umpteenth time, and helped themselves. It was a bucolic scene, but we knew they all had to die. Since both the Bricklayer and Number One Builder still had freezers stocked with the results of two rabbit-assassination sessions we needed to find a new killer, so we invited my friend David to stay, because I knew he was a very good shot.

We'd met as teenaged inmates at an allegedly progressive boarding school. The place applied sometimes-foul vegetarian cooking and liberal good intentions to a disparate collection of boys and girls from self-disciplined strivers to 'challenging' children who were almost ricocheting off the walls. Some really loved it and blossomed, but others were derailed by the place's lack of boundaries. There was a curious, naïve sense of resentment from some of the adults (who failed to be in charge of us) that we weren't making good use of the things that were on offer, but no sense of having some responsibility for this state of affairs.

David and I more or less fell into the latter category, and often ran wild, but he at least had the charm and social skills to make good use of the hedonistic pleasures of co-educational education. I didn't and spent the first four years wanting to go home, and the last year engaged in education-free nefariousness (for the record, it wasn't just me who put the cook's Mini on the cricket square: David helped). This resulted in my not being expelled exactly, but asked not to come back, and of the two of us, David had the most fun. He also had an illicit air gun, hidden under his bed.

There were still traces of the school's Edwardian past in an obsession with fresh air. These manifested in a collective early-morning walk before breakfast, draughty dormitories, turning off the heating at night and – this being the seventies, when the fear of predatory adults was usually unspoken – chalet-style huts for some of the older boys to sleep in, which didn't have door locks. In the winter these lapwood single-skin buildings could be hideously cold, especially since those of us who were banished to them were made to switch off our feeble electric heaters at bedtime. Sometimes ice had to be scraped off the windows – on the inside.

The school wasn't all that far from a dismal, sinister sixties' housing estate, briefly made infamous by a woman who murdered her cross-dressing husband, put his corpse in the back of her car, and dropped it off a motorway bridge. The locals were tough, and relations between them and this apparently fey boarding school were sometimes tense. Occasionally, for entertainment, a posse of estate teenagers would raid the huts in the small hours to hear their occupants scream prettily. Then they made the mistake of doing this to the one that contained David, who had a strong left hook, a short fuse and an air gun.

As they ran away from what had been an unexpectedly painful encounter, they were well lit by sodium street lamps. David extracted his air rifle from under the bed, gave chase, aiming at the general direction of one lad's retreating buttocks, and shot at them. There was a gratifying yelp, and about fifteen minutes later the police arrived. By this time the air gun had been secreted elsewhere and David was back in bed, feigning sleep, the walls of his hut bathed with the rhythmic blue flash of a police car's lamp.

The following day the housemaster asked sternly about illicit air guns. Did anybody have one? This was collectively, and hotly, denied.

'Of course it's ridiculous,' said the housemaster in an aggrieved tone. 'I told them we don't have things like that here.'

And that was that, but for quite some time the school's hut dwellers were able to freeze in private, as the local marauders stayed away.

Since then David has improved his aim on more legitimate targets. As both he and his wife Helen are keen cooks, capable of doing interesting things with rabbit meat, he was happy to act as our third *lapin* Grim Reaper. There was just one problem. When they'd come to stay I'd let the chickens out, and as David hunkered down and prepared to dispatch one of our uninvited guests, rotund and feathery things kept filling his telescopic sight.

'I thought, "If I take out a chicken you'll probably shoot me",' he said afterwards, having abandoned his death watch for fear of dispatching one of our pets; but the following morning David was back and the chickens remained indoors. In due course a cracking retort indicated that the first rabbit had made a permanent career change from cute to culinary. It would not be the last.

BREAKFAST SERVICE

After David, Helen and their bag of dispatched bunnies departed with the cheerful words 'If you want to help us eat your little friends, pay us a visit' the garden was once again the sole preserve of our domestic birds. It was high summer and, rather like our cockerel, Bombay the duck's bits had definitely warmed up with the weather, and he was keen to put them to good use. This was particularly true first thing in the morning. There are priorities, however. Breakfast *before* bonking was his rule of etiquette.

Having shovelled down large quantities of Floating Discs, Crispy was perfectly happy to oblige her better half, a characteristic that set her apart from most of our lady chickens. They would run a mile from an unwanted coupling, because usually it hurt. Perhaps not being cannoned into the ground and having spurs stuck in her sides was an incentive for Crispy, who would take the hint (in the form of a lot of jiggling up and down from Bombay) and lie flat in the grass like a draft excluder with a beak. Bombay would make his presence felt at the other end, and give her what can only be described as 'a good cantilevering'.

If you're ancient enough you might remember a novelty adult toy from the seventies (not *that* sort of adult toy), which was a glass tumbler filled with a blue liquid. Next to this was an ostrich-like glass bird with a bulb-shaped behind. It was hinged at the hips, so that the beak, covered in a sort of felt, could be dipped into the tumbler. Afterwards the thing could be released, when it would spring upright, then flop back into the glass once more, as if it was drinking. It would do this for hours – yes, this was the acme of seventies' entertainment – and the movement was quite reminiscent of a male Indian Runner duck on the job.

So having completed this activity (which might also be seen as a credible impersonation of someone using the handle of a Victorian village water pump) Bombay disengaged from Crispy, who would remain prone for a few seconds before getting to her feet, shaking herself down and wandering off.

As far as Bombay was concerned his latest amorous triumph was best celebrated by a victory run round the chickens' galvanized drinker, as they looked on. It took me a while to realize that his penis often hadn't retracted during these excursions. As the summer progressed and the ground became very hard I could barely watch. Bombay's manhood looked like a length of sausage skin without the meat content, and it would bang along the all-too-solid earth as the duck's outsized webbed feet flapped up and down on either side of it. It says something about pain thresholds that he seemed completely indifferent to the battering this allegedly sensitive appendage was getting. I had visions of it tangling up with his great big feet, causing him to pitch forward into the grass. It never happened, and having regularly seen this scenario played out I would like to stress that I am not a duck voyeur: I just happened to be there. (I'm still slightly surprised that under some circumstances a duck's eyes don't water.)

The garden was now buzzing with avian testosterone, and the ducks and chickens really weren't getting on. Svenson had begun to take notice of Bombay – and not in a friendly way – regularly puffing himself up and stomping after the bird, forcing him to back off. Once Bombay tried standing up to the cockerel, which resulted in his being jumped up and down on two or three times, and having a couple of neck feathers forcibly yanked out.

Things grew worse when Bombay started chasing Bonny the chicken. To begin with we wondered if this was a food-related thing. Perhaps he was making sure that Crispy was getting more than her share by harassing the competition. How wrong we were. Bombay had frankly pervy designs on Bonny. She wasn't a duck, but that didn't stop her being a duck's sex object.

This made Bonny increasingly alarmed, Svenson ever more furious, and Bombay very frustrated. Crispy seemed entirely indifferent to her previously monogamous partner's new role as a deviant love cheat. To start with we thought this was funny, which was a mistake, because things became increasingly frantic and rancorous, with Bonny ever more unhappy as the duck would not leave her alone. We realized things were turning nasty when Bombay trapped and attempted to mount Bonny in the fruit cage, something the bird clearly found very distressing. The situation wasn't helped by Svenson, who piled in, so that his quarry was left struggling under two fighting males.

After that we confined the ducks in the area we'd fenced off for them, since they'd continued to ignore the pond. Peace, of a sort, resulted, with the masculine combatants now sparring with each other from opposite sides of the fence.

We emailed a newspaper Animal Agony Uncle – yes, there really is such a person. What would he suggest? Rather surprisingly, he got in touch almost at once, but what he said did not lift the spirits.

'This sort of thing is common with Indian Runners during the summer. It might calm down, but if it doesn't I'd suggest keeping the birds apart,' said the Animal Agony Uncle, adding rather ominously, 'we've had Indian Runners, and one year we had to lock up the rabbits.'

HEAVY IMPACTS

Whilst all this was going on, Brahms, the creaking door of our chicken flock, seemed to be dying again. She'd stopped eating and drinking, even when we made sure she had access to food and water. The bird looked especially mournful and withdrawn, was getting thinner, and the reason wasn't hard to find. In her crop at the base of her neck, where food is broken down (and then processed in the gizzard), was a lump of something that felt both gooey and unyielding.

'She's got an impacted crop,' said the vet.

Basically, her digestive system had become gummed up, and it was something that could well finish her off. A similar problem had done for Bonny's stepsister Bella, the grey Araucana hatched by Ann Summers.

Birds that are fed too much bread can suffer from this problem, as can hens that poke about on freshly mown lawns, where too many long grass strands can contribute to similar blockages. This was the likely cause of Brahms' malaise. I gave her antibiotics and massaged her crop, but she continued to fade. I decided to mix a little olive oil with her medication, which had to be squirted down her throat with a syringe. By now the bird had been confined to a separate run, where she sat dejectedly until she saw me coming, then tried hiding.

Having caught her, crowbarring apart her hooked beak to administer this cocktail proved difficult as she wriggled, twisted her head and resisted. Eventually persistence and force saw it prised open, and the unguent was slipped in. After that I massaged the doughy ball in her crop, and finished off by squirting down a little water to keep her hydrated. I had to take care: too much water ran the risk of bursting the poor creature's crop, with terminal results. In between these medial interventions I would scoop up Brahms for further neck massages to try and break up the stuff that was stopping her eating. After a couple of days she seemed little better, but I persevered.

*****

In the dovecote the latch-key baby and its grounded parents were doing well. Their child changed from weirdly repulsive to gawky and adolescent. By this time their accommodation was getting distinctly cramped. After one morning's crop massaging I opened the door of their prison to be rewarded by the parents spilling out and making for the summerhouse roof, ignoring the piteous squeaks of their baby as they went. If it wanted to join them it would have to follow. Eventually it tried, having spent ages teetering on the edge of the prison entrance, almost launching itself then changing its mind. On the roof of the summerhouse its parents, still distinguished by fading red splodges on their wings, scuttled about and looked agitated. More teetering, squeaking and flapping followed; then with a lurch the young dove dropped from its vantage point, and flapped wildly, enough to break its fall as it plummeted to the ground, landing in an uncoordinated heap.

Several of the chickens had seen this, formed a posse and made for the small, dishevelled object that sat blinking in the long grass. I spent five minutes chasing them off and hoping the dove would have another go at flying, but the bird had clearly inherited its parent's common sense genes when it came to sitting on the ground and waiting for bigger and nastier things to arrive and give it a kicking. In the end I was forced to try sneaking up behind it, which resulted in the now-familiar sideways dove lurch as it tried to escape, before I scooped it up and plonked it on the summerhouse roof. It scuttled after its parents, squeaking as it went. By mid-afternoon it had managed to fly from the summerhouse to the top of the fence, where it remained for hours, an ideal target for a passing hawk. As dusk fell that's where it remained.

'Here we go,' I thought, inching along the fence, illuminated by moonlight, to try and catch the bird and put it back with its parents (something which turned out to be remarkably easy, since it appeared to be completely transfixed). Fortunately, within about three days it started to get properly airborne. If its flying technique was a little erratic it did seem to be improving, and it was soon following its parents in over-garden sorties.

Jane and I looked on as the little bird copied them as they fed on the lawn and fled charges from the chickens. Every morning it would appear to have matured noticeably. Soon its beak lost its half-finished, plasticine look, and became a small, neat triangle, smaller than the beaks of the other doves, giving it a distinct, snub-nosed appearance. Eventually, this was about the only thing that really set it apart from its peers – that and the juvenile noises it still made. And in this respect it was not alone: the other two pairs of doves had also become parents, their babies emerging into the world looking as tiny, fragile and prehistoric as it once had.

'That was worth all the trauma and early mornings,' said Jane. As we watched the three doves sunbathing on the summerhouse roof, I had to agree.

FENCING LESSONS

Jane and I are a team. We discuss things and work together. Big decisions are jointly considered, so that one party does not do something unilaterally. Abandoning this rule will result in discord. You know the sort of thing; 'Well, I thought you'd like a ninety-inch, flat-screen TV in the living room . . . What's the matter now?'

This should make working together on practical things, like putting up fencing round a duck pond, a relative breeze of synchronized thought and DIY karma. The reality is a little more fraught, as we proved when my venerable hatchback, laden with bales of fencing and wooden posts, hauled a trailer filled with more of the same up our drive and staggered to a halt, feeling only slightly less hot and bothered than the people inside it.

Jane will deny this, but she is rather elegant, but this side of her character had been put to the test in the car, thanks to one of the hefty, round fence posts. This jutted at head height between the front seats, and on a particularly tight bend had decided to get to know her better by rolling sideways.

The result had been a stony fifteen minutes with her head almost pressed against the passenger door window by a big lump of wood as she grappled to prevent it from rolling the other way and concussing me, which might not have improved my driving. Conversation had been reduced to muttered questions along the lines of 'Are you all right?'

'It depends what you mean by all right. I'm no less all right than I was when you last asked me.'

Jane had to stare straight ahead so as not to get a face full of post and risk having to extract splinters from her lips. Our drive goes up a steep incline, and although I was grateful that the post hadn't slid backwards and smashed through the rear window, I was aware that things were not going well. Being constricted into a corner meant that Jane almost erupted from the car like a champagne cork. As I climbed out nerves and cartilage in my lower back tied themselves in a knot.

'Oh bother,' I said. Actually, this is a lie. What I said was far more expressive, and caused Jane to make 'shushing' noises. Our neighbours are a forgiving lot, but Joyce, the lady who lives to one side of us, is old enough to have been around during the Last War, and to have had brothers who fought in it. This would probably guarantee exposure to the riper forms of the English language from an early age, but Jane felt that Joyce did not need to be reminded of them.

My back locked up again, and I clutched the roof of the car, showing the stoicism for which men are famed. Jane gave me a look that said, 'I know it hurts, but if you keep pulling that face I will be forced to laugh.'

'It's not bloody funny,' I hissed. Jane didn't bother to argue, but adopted the tactic of not answering directly, so I knew very well that it was, which was even more irritating.

'I'll get your truss,' she said, striding purposefully into the house.

Bits of your body that don't work, or work less well than they did, are a feature of the human condition, as are vaguely undignified accoutrements designed to help them, like my elasticated 'sports' truss. This is a black girdle of stretchy man-made fibres, with a hilariously inappropriate motif of a stylized running figure. It is held together with Velcro fastenings that always seem to make direct, partial contact with me, resulting in spare-tyre truss rash. This is not a good look, but once Jane had strapped me in I was ready for fencing madness.

We were going to imprison the ducks near their pond, and erecting the stuff we'd bought was a means of doing this.

What followed was protracted, passively bad tempered, and involved uncoordinated manhandling of posts and rolls of fencing made with stout wire and wobbly lengths of wood that would eventually weather and, we hoped, look rather nice. I've never been a great team player – even when that team only involves two people – so there were irritable exchanges about where we should best put the posts.

'I'm not banging them in,' said Jane as I massaged my back. All the while the hens and ducks kept up a constant *'Let us out!'* gabble from their respective prisons.

We have a large, heavy mallet with a handle that's about three feet long, which was ideal for post-battering purposes, but the posts themselves were so high it was impossible to swing the thing down onto them. I had to stand on a rather wobbly stepladder, which Jane kept upright by putting one foot on its bottom rung. Clasping a post with one hand she held another piece of wood over the top of this so that the post itself wouldn't split when I hit it.

Despite being firmly trussed and high on painkillers I found swinging the giant mallet was less than comfortable. With every blow Jane would try not to flinch and duck, although this was unavoidable when shards of timber splintered from the 'protective' piece of wood and whistled past her left ear. Eventually the requisite number of posts were upright, and more or less where we wanted them.

We'd achieved all this without Jane needing a visit to A&E due to a timber-related eye injury, or my ending up in traction. Feeling moderately encouraged we pressed on.

It was dusk when everything was finished. We had a fenced-off pond with a gate and a duckhouse. Bombay and Crispy were all that was now needed, but persuading them to enter their new territory required extended herding and cursing. After much running around they were cornered, caught, and posted into the new duckhouse for the night. By then the slight DIY-related frost that had settled on Jane and I had thawed, as we surveyed our handiwork by torchlight.

*****

Despite my best efforts, Brahms's impacted crop had barely improved. I'd taken to pulverizing some tinned sweetcorn, and syringing small quantities into her beak as a way of keeping her hydrated and to provide nutrients that might find their way past the bulge in her throat which, despite appearing to break up when massaged, always congealed again afterwards. On the fifth morning when I went to get her up I was half expecting to find that she'd died in the night, as she'd looked so miserable the evening before. Instead I discovered the bird standing up in the run outside her bedding area, tentatively pecking and scratching at the grass.

I felt the gooey lump, still very much there, but appreciably smaller than before. Pouring a little sweetcorn into a dish I proffered it to Brahms, who after the briefest hesitation started to eat. Taking the syringe filled with that morning's antibiotics and olive oil, I squirted this onto the food. Why force it down the bird's throat when she could take it herself? Brahms barely missed a beat as she pecked at this, and when I came back ten minutes later, the bowl was empty. To my surprise and delight, instead of keeling over, she'd apparently turned the corner and was getting better fast.

Having dealt with Brahms, I released Bombay and Crispy into their new, fenced-off territory. It took them very little time to work out that they could squeeze their way out again between the bars of the fence and entertain themselves by attempting to bully or shag the chickens. We were alerted to this by terrible screams from Bonny, who was running frantically across the garden, with Bombay literally in tow.

He had clasped her neck feathers with his beak and was attempting to mount her as they ran, which might have been seen as an example of multi-tasking, but Bonny wasn't impressed. So Jane and I spent another protracted period rounding up and imprisoning the hens. This seemed especially unfair, but despite the previous day's efforts we had nowhere to corral the ducks as we'd dismantled their old prison, thinking it was no longer needed.

After that I climbed into the car and headed for the nearest DIY megastore. Returning with a long bale of chicken wire I spent a happy couple of hours with a high-powered stapler, attaching it to the inside of the new fencing from ground to duck eye level. At one point this involved screaming very nearly as loudly as Bonny when she was being molested, because I'd managed to staple my thumb. After that, with a throbbing digit and a strong feeling of ill will, I went in search of Bombay and Crispy, so that they could be reincarcerated, hopefully on a permanent basis.

Once the ducks were imprisoned, I let out the chickens, who more than deserved their R&R. By now Svenson had decided that Bombay needed putting in his place, so he strutted up to the duck enclosure, and stuffed his red face as close to the new wire as he could. This had the desired result of Bombay scuttling up and eyeballing his love rival. Soon the pair were having a serious go at one another through the wire, in a way that was both constant and useless.

Bombay could get his beak through the gaps, but then found it impossible to open it to any degree, making snapping at Svenson a hopeless occupation. This did not stop him trying. Naturally, Svenson saw this as a challenge, and entertained himself by repeatedly charging the duck, but his head was too big to go through the wire and he crashed against it in a macho and futile gesture. I left them to it for half an hour and returned to find them still winding each other up, with Svenson's girlfriends ignoring the sparring, and pottering about the garden.

Meanwhile Crispy was having a quiet swim. Excellent. Both she and Bombay had studiously ignored their new pond. Even chucking bits of bread into it had been met with complete disinterest. Now, finally, one of them had worked out what it was for. Once her partner had given up on his hard man act, I expected him to follow her.

BOARDER DISPUTES

Chickens are incredibly nosy. If you're working in the garden and they're out and about too, you will quickly have a feathered audience. Planting things can be a risky business as a result. Turn your back on a tray of seedlings and you can guarantee that some of the birds will soon be enthusiastically digging them out.

In our previous house the back door opened straight into the garden and the chickens. If we forgot to close it they'd pay us a visit, leaving steaming calling cards to let us know they'd been in. So it was with the area we'd fenced round the pond. This was intended for the exclusive use of the ducks, but the hens would often look covetously at this forbidden territory with the big water-filled hole in the middle.

Sometimes when I opened the gate to feed the ducks, one of the more agile hens would squeeze in behind me. Ann, Too, Slasher and Ulrika proved the most adept at this, and chaos usually resulted. Having bonded with their pond the ducks quickly began regarding the area round it as their own, resented the presence of these chickens, and since Bombay didn't fancy them, he'd chase them off.

This left me with the difficulty of opening the gate to chase them out without encouraging any of the others in, whilst stopping the ducks from hurtling into the main garden. This happened more than once, and when it did, Bombay would immediately rush after Bonny, which terrified her and enraged the cockerel. Once I found myself on the end of a Benny Hill-style chase involving a chicken, followed by a duck, followed by a cockerel, followed by me.

On another occasion I had managed to extract Ulrika from the pond area by chasing her and throwing duck food in the opposite direction. Gratifyingly, the ducks went for the food as Ulrika shot through the open gate, and I pursued her to the far end of the garden, ran back and shut it. 'Job done,' I thought.

An hour later I went into the garden for a breather. I had the place to myself because the builders were away collecting supplies, and although we were still getting along fine, not having to share my home with other people was a great pleasure (although in a funny way I was looking forward to their return).

It was a high summer's day, with pleasant, dry heat rising from the ground. I stood and listened to the wild birds shouting at each other, the buzzing of insects and the ducks splashing about in their pond. Having finally decided to use it Bombay and Crispy had become converts to waterborne life. They loved beating their wings on the surface when preening, and would become completely engrossed.

As I listened it dawned on me that what I was hearing sounded a little different from usual, and that the rhythmic slapping of wings against water appeared more frantic than excited. I began walking towards the pond and as I did so I heard avian screaming. The noise was almost familiar, but the intensity of it was not.

The duck pond lies beyond the lattice fence that by now was partially engulfed by a climbing rose. This was in full flower, obscuring most of what lay beyond it, but there was enough of a gap to allow me to see a patch of churning, foaming water. I could also see Bombay, and underneath him something white.

Bonny. I realized that she must have sneaked in to the pond area when I was chasing Ulrika, and hidden in the bushes, with the intention of hunting for food when I'd gone, but of course I'd shut her in. Bombay had spotted her, and now she couldn't get out and couldn't get away.

As the shrieking and flapping continued I ran to the pond and could see the two birds struggling in the water. Bombay had dragged Bonny into the pond and was trying to mount her as he would a female duck. The chicken's head and neck were visible, but with the drake on her back the rest of her was submerged. He held the back of her head in his beak and kept forcing it under the water.

He'd recently taken to doing this with Crispy first thing in the morning, but she was engineered to put up with such treatment. Bonny wasn't. Luckily they were close enough to the pond's edge for me not to have to wade in and separate them. Instead I reached across and cuffed the deviant duck out of the way. Bonny bobbed to the surface, and I held her clear of the water. She was shivering violently as the commotion died down, then shook her head and sneezed muddy pond water onto my white shirt.

I gave Bonny a visual once-over. Her breath still rattled, but otherwise the bedraggled bird seemed unharmed, if still traumatized. Still clutching her I trudged back to the house, found an old towel we keep for the dog and gave her a rub down, which she did little to resist. Bonny was still damp and cold but I reasoned the summer sun would finish drying her off so I released her with the other hens, who crowded round, and the flock made off in a single, protective mass.

It was only chance that had saved her. The builders hadn't been around, and increasingly were working indoors so they might not have heard the commotion either. Had I come into the garden fifteen minutes later it would probably have been too late. This made me think about Crispy. Would she be the subject of constant harassment from Bombay in the absence of his chicken fetish?

Jane's idea of acquiring more ducks floated into my mind. The fenced area round the pond had enough room for two or three more girls, who would, with luck, keep Bombay more than entertained. Since Ann the Welsummer was still determinedly broody, I took a unilateral decision. We had eight of Crispy's eggs, which given Bombay's recent activity were almost certainly fertile. I moved an apoplectic Ann into our chicken birthing box, and gave her six.

She needed little encouragement to hunker down on them, and when I told Jane about the day's events and what I'd done she didn't object.

'Of course they may not hatch out,' said my wife. 'We could still buy a couple of females, couldn't we?'

GENDER ISSUES
When E. Leslie and J. V. Monaco penned their 1926 jazz classic *Masculine Women, Feminine Men* in 1926, little did they realize just how apposite its sensitive lyrics would be decades later to the occupants of a Kentish garden.

'Masculine women, feminine men/Which is the rooster, which is the hen?/It's hard to tell 'em apart today!' goes the opening verse. I don't know about cross-dressing doves but with chickens, when they're little, it's almost impossible to tell whether they're boys or girls. You may have seen cheerless footage of conveyor belts covered in a sea of yellow, newly hatched factory-farm chicks, and people dressed in scrubs that make them look like cut-price surgeons, plucking tiny male birds from the throng passing beneath, and dropping them on to their own conveyer belt of death. Apparently, although a chicken's private bits are concealed, it's possible to squint, Lord Nelson-like, at a chick's rear and tell from its shape what lies beneath. However, there are, as far as I know, fifteen different types of avian *derrière* to consider, so this is not as simple as it sounds. Most factory-farm chicken sexers use differences in feathers, or down, to decide which are chaps and which aren't.

However, this isn't much help to amateurs like Jane and I. Commercial poultry strains have been bred with very specific characteristics to make this possible. Older breeds – the sort of hens that end up with names and lives of luxury in domestic gardens – haven't, as far as I know. Confusion can persist well into adolescence, and all the good advice about being able to tell the sexes from behaviour, and things like the males growing larger combs, is very hit and miss. Birds can be pretty well mature before the avian equivalents of their balls and other bits drop into place, which was why we weren't entirely surprised when Squawks turned out not to be the bird we'd imagined when we'd bought her.

'That's definitely a girl,' said the woman who sold the bird to us, as a black shape was bundled reluctantly into a cardboard box. (This was the summer before our lives were invaded by the doves, Bombay and Crispy, and the builders.)

Chicken breeders are often as distinctive as the animals they sell, something this lady proved. She lived in a curious no man's land just beyond a large, unprepossessing sixties' housing estate, which stopped abruptly at the bottom of a steep hill, beyond which was open countryside. Separating the two was a slightly sinister-looking lane, which at night was lit by a couple of tired yellow street lamps, and overhung with thick foliage from the mature trees which lined it. These were punctuated at intervals by drives which led to large, 'hacienda'-style villas of the sort fibreglass cherubs were created to adorn.

'When you see an old lorry and some caravans in the undergrowth, you've found us,' said the lady selling Squawks.

We stumbled past these delights and came across a bungalow, with a lawn on which were apparently discarded garden power tools, soft toys that looked as if they belonged to a dog, and pens containing chickens. A deep report of woofing indicated that the owner of the soft toys was close by. Something black and large bounded out of the dark. We like dogs, but we didn't know this one, were on its territory, and it was heading our way at speed.

It was almost upon us when a woman's voice cut through the night. It said something like 'Nooeee-ooowee-noowee-ooweee!' It sounded like someone impersonating a randy parrot. Apparently the dog thought it meant 'Heel!' so that's what it did as two ladies, one elderly, the other middle-aged, emerged from the bungalow and made for us.

'This is Mum,' said the younger of the pair, gesticulating to her cohort. 'The chickens are over there.'

We could now see that the dog was a chubby black Labrador with a sleek coat and a tail which thumped enthusiastically on the ground. It looked as vicious as a bag of toffee. As all five of us made for the hens I felt something cold and damp begin investigating my left hand. Gently closing it round the dog's muzzle, I gave the animal's nose a fraternal shake.

All the hens seemed relaxed and looked healthy. Jane asked about a smallish, active-looking bird.

'That's Squawks,' said the woman. 'Hatched her out earlier in the year.'

'Definitely a she?' we asked, and after positive assurances to this effect, money changed hands, and we were soon back in the car, listening to the occasional scrabbling of claws on cardboard.

To start with we kept Squawks in a separate run so that she could be seen for a period by our existing flock before official introductions were made. When this happened a couple of weeks later Svenson seemed to find the bird's presence especially displeasing, and spent a lot of time chasing it away. When we found it thoughtfully straddling Slasher, who was not pleased, we realized why. This Squawks was a boy, but fortunately the vendor swapped him for one of his sisters, the bird we have now, who has inherited his name but not his proclivities.

A year later, as the longest day approached, we faced a similar conundrum, but this time with the ducks. After the Bonny/Bombay debacle, the drake had become carnally demanding with Crispy. Ann Summers was still stewing her duck eggs, but we were still some way from anything hatching out, and ducklings emerging. It would be autumn before they'd be mature enough to join our existing ducks, so we'd bought two additional chocolate-coloured Indian Runners, whom we named Peking and Hoi Sin, as new girlfriends for Bombay.

'They're definitely females,' said the nice lady who sold them to us. This was what we wanted, as it would give Crispy a well-earned break. His lust for Bonny hadn't wavered either, and he would hurl himself at the fencing that separated the two every time she appeared. Hoi Sin and Peking gave him something else to think about, and on that level they were instantly successful, but his relationship to them was otherwise ambiguous. He chased them, but didn't appear keen to jump on them afterwards.

So Crispy continued to suffer serial molestation. One weekday afternoon I wasn't surprised to find her on the receiving end of an enthusiastic aquatic seeing to in the pond, which meant all of her – bar the top of her head – was underwater, and every so often that disappeared too. There was something wrong with this scene, and it took me a few seconds to work out what it was. Bombay was running up and down on the opposite bank making a lot of agitated noise, because it was Hoi Sin who was making the running with Crispy.

Worse was to come. Days later we saw history almost repeat itself, when Jane decided my life would be enhanced by some gardening. Under mild duress I was digging a hole when I heard Jane say, 'That's not right.'

Hauling myself to my feet I saw what she meant. Crispy was once again someone's love interest, but this time that someone was Peking.

'You know what we've done, don't you?' said my wife. Well, of course I knew. 'We thought we'd bought a pair of girlfriends for Bombay, but they're actually boyfriends for Crispy.'

Three rampant drakes and a single female to fight over wasn't what we'd had in mind at all. We'd hoped Hoi Sin and Peking would distract Bombay's attentions from Crispy and give her a break; instead the unwanted attention had tripled. Female ducks have been drowned by groups of marauding males, and there was now the serious possibility of fights between our three, which could turn nasty.

'I'll ring the woman we bought them from,' I said.

Rather than tell us where to get off, she was surprised but understanding. 'Of course I'll take them back,' she said. 'They aren't what I said they were, but I'm really very surprised.'

Could we swap them for a pair of genuine lady ducks? 'Not at the moment, I'm afraid. I don't have any, but I'll give you a refund,' she said, adding, 'by the way, could you hang onto the ones I sold you for just a little bit longer? We had a fox attack and I want to make the garden secure before I take on any more ducks.'

What could we do? Hoi Sin and Peking weren't trying to jump on us, and we'd rather taken to them, so the idea of putting them in harm's way didn't appeal.

We finally bade farewell to Hoi Sin and Peking (who'd ended up living disconsolately behind some temporary fencing, bathing unsatisfactorily in a plastic baby's bath as, just a few feet away, Bombay and Crispy had rediscovered monogamy and the joys of having the pond to themselves). Things seemed rather quiet after they'd gone, but replacements were on the horizon. We'd discovered another couple who bred Indian Runners in stable buildings near their oasthouse home. Their birds had the run of nearby fields and would charge about in long lines to and from a big pond.

The two we chose, one pale and one dark, were still very juvenile, but were also immensely tame and easy to handle. We realized that they were so young that they'd need to be kept apart from our randy drake and his girlfriend until they'd grown a little bigger, and decided to move them into the fenced-off area that had previously housed Hoi Sin and Peking. Their arrival caused a frenzy of lust on Bombay's part. He paraded up and down in front of them before managing to get his head through a gap in the fence and grabbing roughly at one and yanking her like a rag doll.

'The fencing needs some work,' observed Jane.

In the end I put up a secondary, sort of *Great Escape* prison-camp-style, fence to keep everyone apart, but visible. Even on the day they arrived, the objects of Bombay's desire seemed unperturbed by this sudden change of scene and our drake's obvious intentions. One of them even fed from my hand.

As time passed, they seemed to grow visibly by the day, and would scuttle hopefully towards us whenever Jane and I passed the duck enclosure. By the end of a week Bombay and Crispy seemed completely used to the new ducks, who, inevitably, became known as Pancake and Noodle.

When the time came for formal introductions, it was plain that Bombay was keen to try it on with Pancake and Noodle, and when we released them and they skittered towards Crispy and Bombay, the drake wasted no time in demonstrating that this time the birds we'd bought met with his manly approval. Crispy arched her neck and had a half-hearted go at chasing them away.

As the song said: 'Knickers and trousers, baggy and wide/Nobody knows who's walking inside/Those masculine women and feminine men!'

We felt we now knew, and when a good friend who finds the goings-on in our garden hugely entertaining, rang up and asked 'How are the perverts?', we were able to say, 'Reformed.'

EGGED ON

As the longest day trickled past and the garden baked in late summer warmth, Ann's daughter Ulrika had followed her mother and became broody. This was almost a week after the older hen had begun incubating the duck eggs, and there were plenty of chicken ones for Ulrika to sit on. Svenson had been hard at work, jumping on almost anything that moved, so we decided that Ulrika could have a crack at parenthood as well, since the cockerel ought to have fertilized a fair few hen's eggs.

Beyond hauling Ulrika off her nest a couple of times a day to drink, feed and crap, and to douse her bedding with louse powder, we left her to her own devices. If the bird had her mother's parental instincts, her eggs would be well looked after. Not far away Ann hunkered down on her larger, greener eggs, and viewed any disturbance with a beady eye. Unwanted human attention was something she couldn't entirely avoid, however, because Jane and I had to make twice-daily pilgrimages with containers of water to wet and turn her eggs.

Apparently duck foetuses can stick to the inside of their shells, and although hens will shuffle them round, keeping them damp is something they don't do. Our visits to Ann and Ulrika often brought the other chickens round to have a look at the grumpy proto-parents, either sitting on their eggs like a pair of feathery teapot cosies or – having been forcibly removed from them – stomping round, muttering, eating and excreting crossly.

Time passed, the sun shone, and both birds stuck to their nests. As week three of Ulrika's confinement arrived we tentatively made our way to her enclosure, hoping to find evidence of hatching; perhaps some shattered eggshell or cheeping sounds and bubbling movements beneath the bird's wing feathers that would indicate chicks flopping about underneath.

Nothing. Ulrika remained stoically on the same number of eggs she'd started with. We knew that the next twenty-four hours would be critical, but since nothing else could be done we headed for Ann's lair, extracted the complaining bird and made another unwelcome discovery. Instead of half a dozen eggs there were five.

With a week to go before her eggs were due to hatch it was a worrying sign, as we knew from past experience. The summer before Bonny had taken the hormonal plunge into broodydom, and we'd tried using her to hatch some ducklings for a friend. Three weeks after she'd begun sitting on the eggs, Bonny trashed one. This wasn't a great surprise; twenty-one days is the time a baby chicken takes to gestate, and if she thinks it's overdue, the mother hen will sometimes help things along.

That's what had happened before, and worse was to come this time. As the days progressed the number of eggs reduced to three. With just four days to go before the surviving eggs were due to hatch, I opened the door of the little henhouse where Bonny was confined to be confronted by an angry chicken and an awful, fetid smell.

Removing her from the nest I saw two eggs and the crumpled remains of a third, blood stained and with traces of downy black feathers adhering to it. The bird had heard the chick moving, and because – as far as she was concerned – it was late, had helped it into the world.

Far from being a late arrival, the creature was a premature birth, and hadn't survived. Since the natural world is nothing if not pragmatic, what remained was full of protein (and anyway, Bonny wasn't about to share her nest with a corpse), so she'd eaten it.

The final irony of this unhappy scene was that the bird left her last two eggs alone, but nothing hatched from them, and soon after she returned to the garden childless and keen to get back to a normal life.

Would history repeat itself with Ann? We returned the next day to find her quintet of eggs still intact but, sadly, so were Ulrika's.

We decided to give her another day, but in reality knew what we'd suspected the day before. None of her eggs was fertile, and the cause of that was crowing and strutting round the garden.

Svenson, apparently, had trouser-related issues. Whether these were connected to his aim or to a lack of functioning seed wasn't clear. His libido was in great shape, and he spent hours practising on his long-suffering harem, so the lack of progeny was, putting it mildly, ironic. Later, when we knew for certain that Ulrika was wasting her time, we separated the bird from her congealing eggs, with inevitable, heckle-raising distress.

Back with the other hens she screamed, lunged at Slasher and Squawks, rushed across the garden, shrieking and flapping her wings, then flung herself into a dust bath, where she spent the next hour rolling and wriggling luxuriantly in the dirt. By the end of the week her eggs were in the bin and Ulrika's hormonal passion for them was all but forgotten.

In the meantime things had not been going well with Ann's clutch. We'd had a false dawn on the egg-breaking front, and in the succeeding days their number was reduced from five to four, then three. So far there had been no sign of foster-parent cannibalism. The eggs that Ann had destroyed hadn't been viable, but we had good cause to fear that if anything was alive in the remaining trio, the odds were against it. So I wasn't surprised to find signs of damage on one of those eggs when visiting the bird shortly before we hoped something might hatch out. I could see a neat, circular indentation in its shell, made by a probing beak. Returning a couple of hours later and shifting the complaining bird I found the small indentation had become a bigger hole but, miraculously, the membrane underneath it was still intact. Unsure what to do next – but pretty convinced that this egg was about to be destroyed – I headed back to the house, feeling that our second chicken-/duck-breeding enterprise had been a mistake.

By the time I walked inside, an idea had half formed in my head. Clasping a roll of self-adhesive address labels I headed back to Ann's pen, tore off a small strip of sticky paper and plugged the hole in the shell with it, then left the (by now very flustered) bird to her own devices.

This had the short-term effect of saving the egg, for when I next saw it and Ann, both were where I'd left them. I'd calculated that we had another forty-eight hours to go before anything might hatch out.

Work took me away from the house until the early afternoon the following day, and when I returned I made straight for the henhouse with a bowl of water. Would Ann have started attacking one of the undamaged eggs?

She hadn't, but something else had happened. The egg with the paper was rocking gently, and an area of shell to one side of the repair had been chipped away. Underneath this was a section of torn, red egg membrane, and beneath that I could see a tiny, round eye.

The owner of the eye wriggled and began to make cheeping noises, which caused Ann – whom I'd ejected from the henhouse – to have hysterics. Just before she rushed back in and stuffed herself onto her nest, something dull pink and shaped like a thumbnail poked through the gap and began chipping at the shell. With a start I realized that this was the duckling's beak. I'd miscalculated, and the hatching was a day earlier than I'd anticipated.

As this was Ann's third stint at parenthood, I had every confidence that she knew what she was doing, and managed to keep away for about three hours. Eventually curiosity got the better of me, and I went for another look, hoping to see the tiny bird successfully hatched out.

Instead, very little had changed. There were strong cheeping sounds from the egg, the hole was bigger, but the membrane had become dry and rubbery. I began to realize that my paper repair, which had probably saved the egg and its tiny contents, was now preventing the baby bird's escape. Donning a pair of rubber gloves (to reduce my scent on the egg), and much to Ann's distress, I gently prised away the last of the paper label, revealing a much bigger hole, then left mother and offspring to their own devices, determined not to interfere again.

I managed another three hours before going back. Although the hole had grown a little more, the duckling was still encased in the shell. Ann hadn't helped it, and the noises it made seemed fainter. Was it getting weaker? If I left it and it died after all the trauma and the DIY egg repair, this would be a huge pity. On the other hand, how would Ann react if I interfered again?

Looking at the still-intact egg, as its tiny cargo looked at me, I made a decision: this was kill or cure. I began breaking open the shell. Ann screeched and puffed out her feathers. I kept working on the egg, which suddenly came apart in my hands. Then from the wreckage, something stringy and damp with a blob on the top of it wobbled upwards – the duckling's neck and head.

I let go of the egg, Ann shot forward, expertly tipped it over, and the last I saw was its ungainly contents being spilled out before vanishing under her wing. Screeching again, Ann lunged at my retreating hand.

'It's up to you now,' I said to the furious chicken.

*****

It was dawn when Jane and I made our way to where the birthing drama had taken place. Would Ann have begun parenting, or had she been snacking on an infant duck because it had been contaminated with my scent? As we approached her run juvenile bird noises indicated the former.

Lifting the lid revealed a determined-looking chicken who had something small and busy moving about under her right wing (appropriate, since the term 'right wing' rather suited Ann). We'd brought along a bag of something called 'chick crumb', which looked like dust, but apparently tastes wonderful if you're a small bird.

Pouring a handful into an old yoghurt pot lid, we put this in front of Ann, who began to eat ravenously, making a noise peculiar to mother hens when they want their offspring to join in. There wasn't an interspecies language barrier, because something round and fluffy half rolled, half staggered from beneath Ann's feathers and into the middle of the food dish, where it began to eat as well, cheeping noisily as it did so.

No longer a bedraggled blob, the little creature looked strong, far from wonky, and very duck-like. As Ann began clucking and pushing the food round with her beak to encourage her unusual offspring, it was clear that the pair had bonded.

'Come on,' said Jane. 'I think we should leave them to it.'

Soon afterwards our most senior chicken and our youngest duck were introduced into wider society. In the wild, baby birds need to be mobile very quickly to survive. Ducklings become waterborne after two or three days, and chicks will be out foraging with their mothers in about the same time. So Ann's charge quickly became ambulant, and both of them were soon in the outside run.

This allowed the other hens to get a good look at the new arrival, as Ann fussed over it and tried to teach it how to forage. This revealed some of the obstacles to parenting by instinct, when the instincts and physiognomy of the thing you're parenting are different. Unlike claws, webbed feet aren't designed to haul bugs and other delicacies out of the soil, nor is a beak shaped like a cinema ice-cream tub's spoon.

Ann would stare hard at her baby, then look at the ground, and go through the motions of foraging for food. In return her baby looked uncomprehending and scuttled off, or had a half-hearted scrape, wandered away, or tried to hide under its mother.

'Shouldn't we try introducing it to water?' asked Jane. 'After all, it is a duck, even if it doesn't know that yet.'

We'd heard of chicken parents of ducklings going bananas when their tiny charges rushed off and flung themselves into the nearest pond, and hen chicks reared by ducks following their web-footed semi-siblings into the water and sinking.

We decided on the honourable compromise of digging a small hole in the pair's run and inserting a silver-foil dish that had once contained moussaka. Ann appeared pleased at having an extra drinking receptacle, brought it to the attention of her duck child, and both paid it regular visits, but the duckling showed no interest at all in getting into the thing, and only drank from it.

Another interspecies difference between ducks and chickens was the speed at which their babies grow. Ducks take longer to hatch out, but mature more rapidly after they've arrived. Within a week Ann's surrogate baby had grown visibly and was quickly changing shape, so that the elongated neck and upright posture of its natural parents was becoming very apparent.

'At the rate that thing's growing, it will be too big for the moussaka dish,' I muttered.

'You're not going to try and help it in, are you?' asked my wife.

This seemed like a very good idea, but not to Ann and her baby. When I shoved it gently into the shallow water Ann went ballistic, and the duckling complained loudly, flapped its wings and extracted itself from what for it was already an avian footbath.

'It looks as if we've got a duck who's frightened of water,' I told my wife, in the process having to confess the reason why I knew. In less than a month the duckling was looking its stepmother in the eye, and had begun growing adult plumage. However, its aversion to water remained.

'Let's hope it's a girl,' said Jane, who was already thinking about future introductions with Bombay, Crispy, Pancake and Noodle.

Mollie the cat, who'd spent the last few months scowling at the builders or hiding in dark corners, had increasingly gone into the garden when they weren't around, and one evening I found her hunched next to Ann's enclosure. She'd adopted a slightly arthritic hunting posture and was studying the duckling through the wire mesh. Both Ann and I felt the cat's interest wasn't healthy.

The bird plonked herself firmly on her baby and made a lot of noise, which Mollie ignored. When I nudged her bony backside with my shoe, she resisted, then reluctantly got to her feet, swore and slunk towards the house, stopping at intervals to look back.

Otherwise, everything seemed to be going swimmingly for our aquaphobic duck, beyond an increasing reluctance to use its legs, as the now gawkily adolescent creature seemed to be spending a lot of time doing what teenagers do best: sitting down.

*****

'We've taken the flashing off the roof of your house, so we can join the extension to it,' said the Boss Builder. We were looking up at the shell of our extension, now topped off with a grey slate roof, which stopped just short of the existing house, leaving a gap. To our untrained eyes the structure just needed windows and the remaining slates laying to be almost finished.

'Not really,' said the Boss Builder. He sounded amused. 'We've hardly even started on the inside yet, and there's a lot of work to do there. By the way,' he added, 'Knocker's moving on soon. We need him on another job, and he's pretty much finished here.'

Well, that was the end of an era. We'd already bade farewell to the bricklayer, Bricklayer, and his air rifle, and without Knocker's glowering presence the house and garden wouldn't be the same.

The Boss Builder seemed to be hesitating slightly. 'A plasterer's coming,' he said. 'I think you'll like *him* because he's a bit of a character, but there is one thing . . .'

'Oh yes?'

'He likes to bring his dog to work.'

'What sort of dog?' I had visions of something large.

'She's a terrier bitch called Polly. A lovely dog. Hoover might fancy her.'

I had my doubts. Hoover had been given the snip shortly after he'd arrived, and had shown little interest in the opposite sex since. In fact, when there were vague stirrings in what remained of his bits, he'd tended to go for the wrong sex and the wrong end, and there had been several episodes that involved the other party peering balefully from beneath Hoover's thrusting hips.

Now that summer was on the wane, these half-hearted encounters were likely to fade as the dog's vaguely remembered hormonal urges died with the season. Polly was unlikely to reawaken them.

By contrast, both Svenson and Bombay appeared keen to preserve what for them had been a summer of love for as long as possible, but we knew the impending change of season would act, both literally and metaphorically, as a bit of a cold shower.

AUTUMN

GETTING LEGLESS

As the baby duck matured its reluctance to use its legs seemed to be getting worse, but little did we realize that another duck with walking problems would soon be taking up a lot of our time.

I was driving back from a day spent working in London when my mobile rang.

'It's Pancake,' said Jane. 'I think she's broken her leg.'

How this had happened was a mystery, but my wife had returned to find the bird hauling herself round with one leg, the other flopping uselessly. Both of them were distressed by the situation, and an attempt at catching Pancake caused her to stagger into the water, where she swam in circles.

By the time I stepped out of the car it was nearly dark. Pancake had reluctantly come out of the pond, and Jane had managed to catch and confine her in the old, Dalek-shaped duckhouse.

We decided to get her to the vet as soon as we could the following morning, reasoning that she was stressed enough as it was. Early the next morning, I got in touch with the vet and made an appointment, fetched the cat carrier, and gingerly opened up the Dalek. The thrashing and thumping noises coming from inside were an indication that the Pancake was highly agitated, a situation made worse by my reaching in and wrapping large, clumsy hands round her dry, twisting body.

Although fully grown, she felt sleek, very young, and hotter than I would have liked.

'Come on, come on. It's all right, come on,' I said, in a useless attempt at soothing. Her right leg kicked as she tried to prise off my fingers with small, hooked claws. The left leg hung down, useless and floppy.

Lowering the bird into the carrier, I secured the lid and covered it with a blanket, and was pleased when the bird fell silent. As I walked away the other three ducks crowded by the fence to watch us go.

'It's not good,' said the vet, as we examined the X-rays of Pancake's leg.

What we saw looked like a splintered twig. 'I'm worried she's damaged some nerves too, because the leg seems completely inert.'

We discussed the options, which included putting the duck down, or trying to splint her leg and see whether the bone would knit and the nerves repair. Although the circumstances were different, the situation reminded me of the fox attack and Lewd the chicken. Once again I was confronted with a young bird that hadn't been with us for very long, was now injured, and had an uncertain future.

What made it worse was that Pancake had been the tamest of our ducks, keen to feed from our hands, and trundled happily after us when we visited. Now she was completely freaked out. Not a good start for a bird that was so young she'd yet to lay an egg.

'Do you think it's worth trying to do something to fix her leg?' I asked the vet.

'It's always worth trying, but you'll have to leave her with us – and if she does come home she'll need to be kept apart from the others. It's possible this happened when the drake tried servicing her, and he'll have another go as soon as he sees her again.'

I thought about the damage revealed on Pancake's X-ray, and decided that Bombay probably wouldn't get another chance.

RAIDING PARTY TIME

In the dovehouse it was business as usual. During the summer the flock's numbers had fluctuated from about four to ten birds. When the Bricklayer had put up scaffolding as the extension's walls rose, they'd used it to perch on and peer down at us. This had given them a perfect vantage point to see anyone coming into the garden from the house to feed either the wild birds or the ducks and hens, or to help themselves to any remnants of the builders' packed lunches.

When that happened a posse of white-feathered opportunists would sweep down from the scaffold to feed. This amused the builders but infuriated the domestic birds, who would fling themselves ineffectually at the dove raiding parties, who simply scuttled or flew out of the way. Long gone were the days when the doves lolled about on the lawn and waited to be assaulted by the hens, and although it was no longer possible to tell whether some of these birds were the original ex-Cuckoo Spit animals or their descendants, they did appear to have learned some common sense.

Mostly they looked identical, but one still stood out: the snub-beaked dove that had been our first hatchling, the bird I'd had to repeatedly rescue and keep warm under a tube heater because its gormless parents kept forgetting it existed. Now this bird was deeply healthy-looking and had paired up with another dove, with whom it appeared to be going steady, and Jane and I had begun to suspect that they were raising a family. Doves breed pretty much the whole year round, so although summer had given way to autumn, this wouldn't put them off.

Dove chicks are often very vocal, and on garden sorties to have a break from writing or involuntary exposure to Radio 2 I would sometimes hear high-pitched squeaking from the dovecote. These were the sounds of its youngest residents demanding food.

I knew where in the dovecote the ex-baby and its significant other was living, but so far hadn't heard the tell-tale chick noises, so one afternoon I climbed up a ladder and peered inside. In the darkness I could see an adult – although I couldn't tell which one – stuffed into an uncomfortable-looking nest made from old bits of twig, feathers and other detritus.

It gave me an unwelcoming look, and I climbed down the ladder, none the wiser as to whether or not it was sitting on a clutch of eggs. Back at the house the scaffold had long been taken down, and the builders were working inside, so the remaining doves had taken to perching on the extension itself. Rather than fulfilling my Dad's prediction that they'd use its walls as a toilet, the birds had instead found a way of crapping on the cars below (or when they weren't there the drive, leaving an off-white residue, which Hoover had taken to licking).

The doves seemed to operate as a unit, and although there was some low-level squabbling there seemed to be a unity among the birds, which boded well for the flock. My train of thought was interrupted by the phone ringing, and I managed to get inside the house, clamber over tools and building materials and pick the thing up before the caller rang off.

'We're not going to charge you for the ice lollies,' said a voice. It was the vet. The problem of finding the ideal splints for a duck's broken leg was answered by a trip to the newsagent and the purchase of some ice lollies for the vet and her staff. Once they'd been eaten, the sticks were fashioned into duck splints, and bound tightly to Pancake's leg. A painkilling injection had given her a relatively comfortable night, and once we were ready to pick her up she could go home.

'I'm still not convinced about the leg, even if the bone knits,' said the vet, 'but the repair's gone better than I'd expected.'

Once again in the corralled area where she and Noodle had lived when we first brought them home, Pancake looked unhappy and pathetic, and I wondered if we'd done the right thing. With her right leg bound tightly to the splints by bright-blue webbing, she found movement hard. If either Jane or I came anywhere near her she became terrified and would half stagger, half hobble as far away from us as possible. This was a very different bird from the one that had rushed up for food. In fact she appeared disinterested in eating, which was another problem, because we were going to have to use food as a medium to give her the liquid antibiotics the vet had prescribed. In the end hunger did get the better of her, and she ate, but not with her previous greedy enthusiasm; this was an entirely functional process. To encourage her we began feeding the others from a bowl on the other side of the fence where she could see them, but she seemed to ignore their obvious pleasure.

Bedtimes were a source of stress too. With the aid of a torch Bombay, Crispy and Noodle would follow the beam and waddle up the ramp into their home, and I'd shut the door on the sound of beaks vacuuming up evening treats. Not so Pancake. I'd try laying tracks of bread to her duckhouse, but the bird was more inter-ested in running away from me than feeding. She couldn't stay outdoors, so night after night I was forced to catch her by torchlight. To begin with this resulted in hysteria, with the bird clumsily flinging herself from corner to corner with such violence that I worried she'd injure herself again.

After a couple of weeks things improved a little. Although Pancake spent most of the day slumped sullenly in the corner of her little run, she no longer actively tried to hide whenever she saw people, and had become a little more enthusiastic at mealtimes. When it came to putting her to bed we'd come to an arrangement which involved my walking gingerly towards the duck. She would fling herself around for a minute or so, then hobble into the Dalek, where she would cringe as I shut the door.

Even if Pancake didn't appreciate it, at least this was progress.

GROWING UP, SITTING DOWN

Our chicken-reared duckling was now appreciably taller than its mother, although both remained devoted to one another. The fact that the duckling was also getting too tall for the run and enclosure it shared with her also seemed to encourage it to sit down even more.

When it walked we noticed that its steps were becoming stiff and uncertain, that its legs seemed to shake if it stood for any length of time, and that it would almost collapse onto its backside after a few minutes. We speculated that its extended confinement in the egg had caused the problem.

'That bird needs some exercise,' said Jane, so we confined the rest of the hens to barracks, and let Ann and her curious baby into the garden for the first time. As we watched them trundle off I told Jane that I'd thought of a name for our garden's newest inhabitant.

My wife gave me an 'Are you about to be silly?' look.

'The thing is, Ann's full name is Ann Summers.'

'Yes,' said Jane. She spoke slowly and deliberately; in the way primary school teachers do when they're about to be cross with small and lavatorial boys.

'What about naming the duck after an Ann Summers' accessory?'

'No,' said Jane.

I opened my mouth to speak.

'No,' she repeated, in an attempt to put a stop to the crudeness she rightly expected to follow.

'Dil . . .'

'NO!'

'. . . do.'

'We are not calling a duck Dildo,' said Jane. 'It's not a good name. It's pervy and sick.'

Which is how we ended up with a duck called Dill, stepchild of Ann the chicken. Two animals with abbreviated names that gave no hint of the *double entendres* lurking just syllables away.

'You are not to tell anyone that bird's full name,' said Jane, which of course prompted me to tell my Dad. He laughed and then told my stepmother, who didn't.

*****

238

Dill remained convinced that he was a chicken. When he wasn't sitting down the bird would waddle after the hens who'd finally met him on neutral territory, and his stepmother, over whom he now towered. He'd found the adult Indian Runners briefly interesting as something to look at, but the excited, frank interest they'd shown him in return alarming, and had steered clear of them since.

By contrast the chickens didn't seem to have a problem with Dill. Ann was the most senior female bird and he was her baby so he was, mostly, treated as one of them, but it was obvious that he wasn't. He really didn't fit in, and there would be an inevitable parting.

We'd noticed something similar with the builders, not because of the arrival of the Plasterer, but the unheralded appearance of the Electrician. Both were distinct characters, but only one of them belonged with the others.

This was the Plasterer, who on first appearance was the less conventional, and hardest to fathom. He drove an elderly grey Transit van. On his lap sat a short-legged terrier bitch with wild, sticking-out fur, who was introduced to us as Polly. These two were well matched, because although the dog was small and the man large and shambling, both seemed to view the world with a satirical eye.

The Plasterer was pear shaped, with a broad, amused face and a shock of black hair that stuck out like his dog's. If Beethoven had gone into the building trade and had done a lot of smiling when having his portrait painted, he would have looked quite like this man.

The other builders greeted him warmly, and after a few days I noticed that everyone was laughing more. The only member of our extended household who was less enamoured with the change was Hoover, and this was thanks to Polly. She fancied him, which our dog found worrying rather than flattering.

Hoover had been given the snip years before, and his sexuality was somewhat ambiguous. In moments of excitement he'd still do that 'thrusting hips' thing, but often at inappropriate moments, such as when meeting friends' small children ('Yes dear, you're right, he *is* dancing').

When Polly arrived he was eight years old, and had largely given up this sort of thing, and didn't seem pleased to be the cause of a great deal of canine cavorting, nuzzling and general flirting from a younger woman. After one particularly unsubtle approach, which had all the hallmarks of a furry version of the film *Basic Instinct*, he jumped onto my lap, showed the whites of his eyes then buried his face in my jacket. Polly danced round my feet and everyone laughed.

'Is your dog a poof?' asked the Plasterer.

I got the feeling that the Electrician wanted to hide as well. Short, neat and of indeterminate middle years, he had a high-pitched voice, a precise, pedantic way of speaking, and like Dill and the chickens, didn't fit in with the others, although it was obvious that he wanted to.

He joined in with the banter, but never seemed to be entirely part of it, and his contributions soon began sounding forced and a little shrill, breaking up rather than blending in with the ebb and flow of conversation. He also had an unfortunate way of explaining what he was doing in impenetrable, tedious detail, but this also sounded as if he was telling the others what to do.

As he ripped up floorboards to lay new cable, shut down the power and generally did things that were at cross purposes with everyone else, the atmosphere changed. The camaraderie between the other builders was still there, but it had been joined by a mutual irritation with the Electrician. They didn't like him, and I found, to my private discomfort, that I didn't like him either.

This was irrational, since the man had done nothing to me personally, and seemed conscientious to an almost absurd degree, but just couldn't help being irritating. I had a discomforting sense of being pulled along by the majority in a primitive way I hadn't experienced since my teens.

So I was quite relieved to get out of the house and take Pancake to the vet.

*****

She was smiling as she showed me a fresh batch of X-rays of Pancake's leg. The patient, who was still nervous of me in the garden, perversely seemed to find the consulting room less frightening. She stood on the examination table and eyed us enquiringly.

'That's knitted really well,' said the vet, indicating a picture of a clean-looking single line of bone. 'I think we should take off the splints.'

Further encouragement could be found with the duck's reflexes. Before taking scissors to the bandage the vet pushed her hand against the bottom of Pancake's foot, which appeared to be offering resistance, rather than flopping about. 'I'm pretty confident that we've got away with this,' said the vet, as she reached for some scissors.

As she got to work on the bandage the duck barely struggled, until the now-dirty webbing was almost cut away and the makeshift splints removed. Suddenly Pancake launched herself at the ceiling, flapping her big wings in the vet's face, and landing cleanly on the floor. She scuttled towards the door, and stood quacking loudly, the remains of the bandage trailing from her foot. 'Do you want to go home now?' the vet asked Pancake.

This was wonderful. The duck's knee joint was working as it should, and she could walk and stand properly.

'When I first saw her I really didn't imagine we'd get to this point,' said the vet, who was clearly as delighted as she was surprised. She suggested another couple of weeks away from the other ducks, before official re-introductions 'to make sure everything's really strong before Mr Duck has his wicked way with her.'

Back in her compound Pancake was still skittish, but now able to move properly. No longer in discomfort, she began regaining her interest in feeding, foraging and her compatriots. She even began to sit tentatively in the water-filled baby's bath we'd buried in the ground to give her somewhere to have a swim.

If anything, Jane and I were now more concerned with Dill, who walked like an old man, on shaky, wobbly legs. His maleness was also becoming obvious, because the noises he was beginning to make and plumage he'd grown were the absolute spit of Bombay, his biological father.

By then we wondered if he needed encouragement to keep moving and to get stronger. Lustfully chasing the chickens wasn't on the agenda as a means of exercise. Unlike Bombay, Dill didn't find Bonny or any of the others attractive in a *'Hi girls!'* sense. Still, we did wonder how our cockerel was going to react to the bio-logical son of his chief nemesis, and had put off finding out by keeping Svenson in dock when Ann, Dill and the other chickens shared the garden.

We found out sooner than planned when Svenson escaped. I'd noticed that the bird's water trough needed filling, and in a move worthy of an American football player, Svenson shot round my boots and exited his aviary before I'd had the chance to close the door. All I could do was watch his feathery posterior as he made a beeline for his girls and a confused, sit-down duck.

Svenson charged Ann, but thought better of it when she hunkered down and fluffed up her feathers in a way that was easy to interpret as *'Sod off!'* The cockerel took the hint, veered sideways and ran straight past Dill – who looked a little surprised – then proceeded to attempt to have his wicked way with Bonny.

This caused her to start running in the opposite direction, with Svenson in hot pursuit. The pair rushed past me and into the henhouse, from which the sounds of a complaint and a serious scuffle could soon be heard.

It took about an hour for Svenson to notice the unusual addition to the flock, who'd been waddling behind his Mum and the rest of the hens as they bore down on the feed bin for their supper. The cockerel was in the middle of pointing his girls in the direction of the best grub when he stopped mid-cluck, straightened up and stared hard at Dill for about five seconds. Then he went back to feeding. Had he been capable of it, I could almost have imagined him shrugging.

BEDTIME STORIES

Despite the duck imposter, things remained pretty calm with the chickens, although Dill's habit of plunging in at feeding times did result in some irritation and the occasional jab from Bonny, Ulrika, and even Slasher.

Svenson too would sometimes give Dill the evil eye, which was usually enough to put him off; but we knew that it couldn't be long before we'd need to introduce him to the ducks, but only after Pancake had returned to the flock full time. She continued to stomp around on her own, peering through the fence of her duck compound at the others. Although much less stressed than before, she remained spooked by Jane or I going anywhere near her, and this neurosis seemed to have rubbed off on Bombay, Crispy and Noodle, particularly at bedtime.

Now, as Pancake made an hysterical vocal contribution in the background, they would swim in noisy circles in the middle of the pond and refuse to come out. To combat this I'd started putting her to bed first, but this too was a fractious process, which caused the others to become even more agitated. Having deposited the now strong and very mobile Pancake in the Dalek, I'd have to endure five or ten minutes of sustained muffled quacking from inside it.

This wouldn't encourage cooperation from the others, who'd stay in the pond and join in. I'd get a piece of wood and churn the water at one end; eventually they'd exit, rush into the duckhouse, then rush out again before I could get to it and close the door.

Over the period of an hour or so I'd make two or three trips to the pond and fruitlessly try and get its occupants to turn in for the night. Eventually Jane would often be drafted in to help herd them to bed, with all of us becoming increasingly stressed and bad-tempered.

Things reached their nadir when the ducks discovered an inaccessible patch of undergrowth behind a wooden arbour seat, which I could only get to by crawling on my hands and knees. After Jane and I had made an attempt at persuading the ducks to go to bed – and my wife had refused to come out yet again – they'd scuttled into their thicket and I'd lost patience.

'Right!' I snarled. 'I'm coming to get you.'

Crawling through over rotting leaves, moss and dank soil, I didn't think what these things were doing to my trousers. Brambles and twigs snagged my shirt as I swore and inched forward, trying to angle my torch at the shrinking, vocal trio of ducks. Finally I lunged and got hold of Bombay.

Clutching the struggling duck and reversing, I became aware of a soft pattering noise and realized that it had started to rain. By the time I'd bundled him into the duckhouse and shut the door, what had begun as a mild shower had turned into a freezing downpour. There was no turning back, so clutching the slippery rubber torch, I again crawled along the now soaked, mulch-like ground, cursed as sodden foliage slapped into my face, extracted and imprisoned Noodle, then went back for Crispy, who resisted to the last.

When at last I slopped back into the house I was caked in filth, soaked and furious. Jane nearly fell off the sofa with surprise, insisted that I had a shower and said she'd fix up a change of clothes. Outside a strong wind was battering rain sideways into the house, as I peeled off my ruined clothes and stepped gladly into the shower.

Halfway through its blissful, hot sluicing I thought I could hear Jane shouting. I shut down the shower and heard the unmistakable sounds of human agitation. Still dripping I ran onto the upstairs landing, which was now dark. Jane was standing pointing a torch at the ceiling, which I could now see was dripping even more than I was.

'Water's pouring in,' she said. 'I've turned the light off. You can see it running down the fitting.'

Rain was also cascading down the walls, making the wallpaper bulge. Why was this happening? Then I remembered that the builders had removed the flashing from the old roof to join the new one to it. This was acting as a culvert and channelling water into our attic.

'We've got to do something.'

'I'll get some buckets.'

'Good idea,' said my wife, 'but put some clothes on first.'

*****

Jane is better at finding life's positives than me. She also has a knack of turning bad situations to her advantage, so when the builders arrived the following morning to survey the soggy landing, she seemed remarkably upbeat.

They weren't, as they huddled, shuffled, and in the case of the Boss Builder, used the word 'sorry' as verbal punctuation. I wasn't enthused either, not only because of the state of our house, but also because I knew it had

247

given my wife an idea, and I had an awful feeling that would mean more short-term chaos.

'The plaster here and in the hall's always been knobbly, and the wallpaper's awful,' she said. There were nods of agreement.

'I don't suppose the soaking has improved it.'

More nodding.

'Would now be a good time to replace it?'

My shoulders slumped. The Boss Builder gave me a sympathetic look, then turned to my wife.

'Well yes, I suppose so,' he said. 'But it will mean a bit more mess.'

PECKING ORDERS

By now the nights were getting noticeably longer, and the leaves on the trees were looking jaded and tired. If the days were still pretty balmy the nights weren't, so the chickens and ducks clumped together for warmth, whilst Dill and Ann continued to sleep in a feathery huddle in the little henhouse where he'd hatched out.

'Before you know it, it'll be Christmas,' I said to Jane, who responded by telling me I was 'a very evil man' for reminding her.

We were in the garden watching the birds. I'd seen a great deal more of them recently as I'd more or less moved out of the house during working hours. Its interior had become a swirling mass of plaster dust and noise, as the hall and landing were stripped. I knew it wasn't the builders' fault that what they'd now been asked to do created 'a bit more mess', but when they were working, I couldn't.

Mollie had taken to hiding under a bed, and Hoover had become my constant companion. He couldn't cope with it either, nor had he got any better at dealing with Polly's bouncing advances. So, hunched in the summerhouse, we'd watched everyone else getting on with life.

Ann and Dill still did everything together, but Ann appeared to have given up trying to teach her gawky duck child how to forage. By now they made an incongruous pair, since Dill was a lot bigger than his foster parent, and looked as if he was old enough to take care of himself.

So I wasn't surprised when over a three-day period Ann's hormones came to the same conclusion. She began wandering off to be with the other hens, and I'd see Dill wobble to his feet and plaintively rush after her. The noises Ann made were also changing. Broody birds screech when disturbed and make distinctive 'clock-clock-clock' sounds the rest of the time.

We were into September before Ann's tone changed from 'clock' to 'cluck'. Then one night I came to put the birds to bed to find Ann back in the main henhouse with the other chickens, and Dill sitting on his own in the little hutch he'd shared with her. The following day Dill's one-time protector blanked him, and the unhappy bird spent the day shadowing a now completely indifferent chicken.

'It's time he came out as a duck,' said Jane.

In the house much the same thing was happening to the Electrician. At mealtimes the others would sit together, but as Hoover made Scotch-egg-goggle eyes at the Male Bimbo Builder, the Electrician sat on his own, reading the paper.

The man's habit of explaining his every action in lugubrious detail seemed to have grown even worse as the atmosphere soured and was, I suspect, a reaction to it that he just couldn't help. Once, I'd ventured into the house for a pee, but before I could return to a particularly awkward piece of work, the Electrician pinned me to a wall and began talking about why he couldn't do something with a light fitting.

'Not now!' I hissed. He looked sad and I instantly felt mortified, but not sufficiently mortified to go back and apologize: my ego wouldn't let me.

I wasn't doing well at rising above the atmosphere of opprobrium, and was very nearly joining in, but the poor sod was so *irritating*. I also realized, with a guilty start, that I wouldn't have reacted in the same way to any of the other builders.

Certainly not to the Plasterer, who had a hard-to-describe charisma, and an often dark, surreal sense of humour. There'd been a murder where the victim had been buried, described in enthusiastic detail on a radio news broadcast. As he mixed a bucket of white gloop, the Plasterer muttered, 'Buried her? Suppose he thought he could grow another one.'

The way he worked was a revelation. Like the

Bricklayer he turned a humdrum-looking job into an art form. I'd watch, fascinated, as he'd layer plaster onto a board, and with huge arc-like movements apply it perfectly to walls and ceilings. There were never ripples, bubbles or undulations. Every casual-looking sweep left a surface that was ready to paint as soon as it was dry. What he did must have been physically hard, especially climbing up ladders and holding the board, laden with plaster, over his head for hour after hour.

'You don't see many old plasterers,' confirmed Number One Builder. 'They get joint problems.'

Apparently this one, who was in his forties, was already a veteran. All the while he talked about anything, from architecture to cookery, in the knowledgeable way of someone who knew stuff because it was interesting rather because he wanted to impress.

He spoke with a roll-up-smoker's rasp and was one of those unconventionally articulate people with a creative flair for profanity that was clever rather than crude. There was also a danger about him, a beneath-the-surface volatility that he shared with Knocker, and I wasn't surprised to discover that they shared digs.

The Plasterer was good company and endlessly interesting, and in the end I had to keep out of his way to avoid stopping both of us working. Unlike the Electrician, I realized that I wanted this man to like me, and that had an effect on how I treated both of them: being nice to one and, if I wasn't careful, bullying the other.

Back in the summerhouse, I watched Ulrika stomp over to Slasher, who was minding her own business, and give her a practised, vicious peck, causing her to run past Ann and Meringue. Those two were standing about, apparently amiably, although Ann – as the bird at the top of the pecking order – could have done the same to any of the others, but they were never going to give her a hard time.

All of them could have bullied Squawks. So could Slasher, and when she saw the bird she did. Had it been Bonny or even Too, the encounter would have been different.

Strip away a few social niceties and that's how things were in the house.

COMING OUT

There wasn't a lot I could do about that in the short term, except try to be nicer, but doing something about Dill's singularity had got a lot easier thanks to Pancake having rejoined Bombay, Crispy and Noodle. We'd opened the fencing of her little enclosure, and she'd rushed out and was soon sharing the pond with the other three.

When I fed them, she ran into the mêlée in a way that indicated a complete recovery from her broken leg. The trauma that went with this had also receded, and within a very short space of time she was again feeding from our hands.

This meant Dill could move into the Dalek and the area we'd fenced round it. I waited until night and put him to bed in his new home, and the following morning he emerged looking more than a little lost. The older Indian Runners found him interesting, peering at him through the fence as he ran up and down looking for a way to rejoin Ann and the other hens, who pottered round the garden, apparently unaware of the stress their presence was causing.

'Look mate,' I said to the scuttling figure, chucking some duck food in his direction. 'Admit it, you're not a chicken.'

It took a while for this fact to start percolating into Dill's brain, and the first manifestation that it had was when I found him floating thoughtfully in the plastic baby bath we'd used for Pancake during her confinement. He was starting to take an interest in the other ducks too, staring at them through the fence.

I looked indulgently at this happy scene, not realizing that the duck-weaning process couldn't have been better timed.

SUMMER'S END

I'd finished feeding the hens when I noticed the still bundle of feathers in the corner of one of the aviaries: a dead chicken. My heart lurched. The plumage was dark, and for a minute I thought the thing had been decapitated, because I couldn't see a head.

After the initial shock I took a closer look. The bird was intact, but it appeared to have pitched forward, so that its head was tucked under its body at a grotesque angle. My first thought was that this was Too, but then I realized that I was looking at Ann Summers.

She'd died just weeks after seeing Dill into the world, had left behind two successfully reared generations of chickens, and spent her final summer at the top of the pecking order. In terms of henhouse achievement, she'd nothing left to prove.

Jane was at work, and there was no room for squeamishness, so I donned rubber gloves and lifted the stiff, lifeless body and checked for signs of injury. I could see none. Ann was about eight years old and had probably suffered a heart attack. She'd been full of beans the day before, eating enthusiastically and dispensing summary justice.

It struck me that she'd probably been extremely healthy until the moment she'd become very, very ill, and this was probably a good end to a successful life. As I contemplated the now-cold object that had lived it, I felt a light pressure on one of my booted feet and looked down to see Too standing on it, head cocked to one side so that she could see round Ann's remains and look me full in the eye.

I'm not saying that she knew what had happened, that here was some elemental communication between man and bird. It was obvious that Too didn't give a bugger about Ann, and wanted to be fed yet again.

Had it been the other way round the only difference would have been that Ann didn't have to stand on people's feet to get their attention.

Too continued to stare.

'In a minute, sweetheart. I've got to dig a hole first.'

*****

As I pulled back the fencing that was keeping Dill apart from the other ducks, and he tentatively waddled after them, it occurred to me that he'd probably forgotten about Ann too, which was exactly as it should be. She'd certainly forgotten about him before her final exit.

Dill had other things on his mind: chiefly how to assimilate with four ducks that viewed his presence with interest but some caution. The bird would toddle in their direction and stop. Then they would run towards him as a group. He'd run away, as they followed in a line, like a tiny fairground ride. Sometimes they'd stay put and stare at each other until Bombay decided to assert his authority, arch his neck and then hiss at the new arrival.

We weren't sure how things would work out, but reckoned two male and three female ducks ought to be able to get along, and after a few days of being semi-chased away, Dill became a fully paid-up member of the flock. Both the male ducks appeared to be fine in each other's company, and as Dill filled out the only way to tell him apart from Bombay was that his beak had slightly different colouring.

'Have you noticed that he's no longer wobbly on his pins?' said Jane.

She was right. Dill had become extremely mobile and strapping. All traces of the tremor in his legs were gone, and he was soon putting his newfound strength to good use as he pursued the three females.

We wondered if Bombay might decide to express his displeasure, suspecting it was only a matter of time before he did, and weren't surprised when the pair ultimately fell out. It was evening when we found one male duck dragging the other about by grabbing beakfuls of wing and neck feathers. Then we realized that it was Dill who was the aggressor. The one-time cute duckling was now very grown up and keen to fulfil his genetic destiny by deposing his old man.

Which is why Dill found himself in a box in the back of our car and heading for the duck breeder who'd sold us Pancake and Noodle.

'We've got plenty of space, plenty of girls, and if your drake tries anything silly, ours will soon put him right. They won't stand any nonsense.'

We'd become very fond of Dill, but he was one boy too many. Still, as we carried the box and its scrabbling contents to the edge of lake near where we'd first seen Pancake and Noodle, we did wonder how he'd cope with the change of scene. We put the box on the ground and dithered until the duck breeder gently insisted that we got on with it.

Dill spilled out, saw a pond full of web-footed totty, and making noises that were clearly his equivalent of 'Phwaaaaa!' rushed to join them without a backward glance. We left him intent on making as many conquests as possible, in between being chased off by a bunch of irate drakes.

FINISHING OFF

Although we'd had a couple of departures, there had been two arrivals as well, in the shape of a pair of dove chicks. They were the products of the ex-baby dove and its partner, although we never did discover which was the mother and which the father. I'd first seen them a day or so after Ann's demise, when both their parents had been away feeding on the bird table. I'd sneaked up a ladder and peered into the dovecote, where it was possible to see two dove squeakers, a bit older than the one-time baby when I'd first clapped eyes on it, but still fabulously ugly. They sat motionless in the lumpy-looking twig nest and pretended I wasn't there.

Nothing malign discovered them afterwards, and by the time the builders had fixed the roof and turned our extension from a brick- and breeze-block shell into something approaching a finished part of the home they'd fledged, and had joined their parents and the other doves on feeding expeditions.

When they got to that stage I looked at them through the French windows of the extension, ran my hand along its glass-smooth plasterwork, and thought about the hall and landing, which had received similar ministrations. Even I could see that what had been done there was a beautiful job too, which I could now appreciate since my self-imposed summerhouse exile had also finished.

During my absence, nobody had murdered the Electrician who, pedantic to the last, had virtually rewired the house and had done so in a way that every-thing worked without a hitch. He would be coming back just once more, with a man from the council, to sign off his work as fit for purpose (having first told him at length exactly what had gone into it). It would actually be good to see him, because he'd worked hard and I wanted to say thank you, but I knew it would be even better to see him finally go.

The other builders were coming back to finish the job, which would take about another three weeks. Although Jane and I were looking forward to having the house to ourselves, they'd been good company; we'd never exchanged a cross word and I knew the place would seem very quiet without them.

We'd already said goodbye to the Plasterer and Polly, much to Hoover's relief. Later we heard that the little flirt had parented a litter of puppies, one of whom the Plasterer had kept because it had a hole in the heart. It was called Esther, apparently because he liked taking both dogs to the pub and shouting 'Polyester!'

GARDEN PARTY
It was a Saturday afternoon and this time my hand wasn't stuck, although winter was imminent and I'd wrapped it round something cold.

258

The clocks were soon to go back, trees were swathed in autumnal yellows and browns, and leaves were falling. Under our elderly pear tree Too was busily raking through piles of them that I'd swept up. She looked very businesslike as she heedlessly vandalized my work.

Soon harsher weather would finish stripping the trees, but that was still to come, and it was unseasonably sunny.

'Drink in the garden?' asked Jane.

The idea didn't seem so daft, and since it would probably be the last opportunity we'd have to enjoy this small piece of decadence until the following spring, we wasted little time in setting things up. Wicker chairs and a table were brought from the summerhouse to the middle of the lawn. Then Jane and I, swathed in outdoor clothes, sat down with a couple of glasses of red.

Hoover arrived and planted himself on Jane. Too abandoned 'her' leaves and came over, and soon a posse of chickens had surrounded us. We'd brought some crisps and a small bag of grain for the hens. Hoover, who was sitting like the HMV dog on Jane's lap, likes crisps, so took turns with the birds for regular refuelling.

Trailing a withered bramble branch, Squawks rushed up and tore a crisp from Jane's hand. Too jumped onto my lap, pecked at one of my jacket buttons then flapped indignantly as I pushed her off again.

'Look what she's done to your trousers,' said Jane. I did and saw splayed and smudged earthy claw prints on their off-white fabric.

'You won't want to see the underside of my left shoe then.'

'Why? What have you trodden in?'

'I'm pretty certain it's rabbit droppings.'

The ducks were outraged (although not about my trousers, or the return of the rabbits). They could see the hens being fed, but were unable to join in. Putting down my glass, and brushing my lap (which had the effect of mashing soil into my trousers), I walked over to the ducks' enclosure, reached over the fence and lifted the lid of their food bin. Inside I found a bread roll, which I broke up and threw for them so that they wouldn't miss out. Birds may not have the biggest brains, but they possess a definite sense of fairness, as in: *They've got some, why haven't we?'*

Soon we were joined by the doves and watched as the former baby, its partner and their offspring swooped in to help themselves to some of the ducks' bread. Despite the lateness of the year, I could still feel the sun's warmth on the back of my neck as I turned and accidentally kicked Brahms, who'd made a discreet appearance near my left foot. She didn't run away, but instead lowered her head and stayed stolidly where she was.

'Sorry, sweetheart,' I said, offering a piece of bread roll, almost at ground level.

After hesitating, the bird took it then lumbered for the nearest bush, so that she could eat unmolested by the other chickens. Hoover was now curled up like a furry seashell on Jane's lap. The dog was almost asleep, eyelids like drooping half moons.

Svenson, who was about two feet away, stood on his toes, stretched, flashed his orange-rubber spur covers and crowed lustily, causing the dog to jump. Jane squeaked as the edge of the glass she'd been in the process of putting to her lips collided with her front teeth.

'Thank you,' she said, as Hoover relaxed, tail thumping against the sleeve of her jacket.

'You all right?' I asked, as Jane pretended to study the bottom of her glass to see if it contained bits of enamel. 'Very much so,' she said, as the hens milled about us and continued to look hopeful.

Jane looked at our bigger, waterproof, almost completed house, which was no longer a repository for packing cases, dust and testosterone, only a deaf old cat, asleep on a favourite, fetid cushion. A dove swooped overhead, returning to the dovecote with a piece of bread it had stolen from the ducks, who were left running around and shouting at it.

I took a draught of wine, felt the alcoholic warmth gently spreading through my body, and said something to the effect that I couldn't think of anything that would improve the moment.

'Well,' said Jane. 'What about some geese?'